The Year Science Changed Everything

1957's International Geophysical Year and the Future of Our Planet

MARK O'CONNELL

 Prometheus Books

Essex, Connecticut

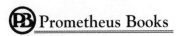 **Prometheus Books**

An imprint of The Globe Pequot Publishing Group, Inc.
64 South Main Street
Essex, CT 06426
www.globepequot.com

Distributed by NATIONAL BOOK NETWORK

British Library Cataloguing in Publication Information available

Library of Congress Cataloging-in-Publication Data

ISBN 9781493084906 (cloth)
ISBN 9781493084913 (electronic)

♾™ The paper used in this publication meets the minimum requirements of American National Standard for Information Sciences—Permanence of Paper for Printed Library Materials, ANSI/NISO Z39.48-1992.

To my dad, John E. "Gene" O'Connell,
for instilling in me a love of travel, adventure, and learning

Contents

Prologue: Half-Remembered   vii

Acknowledgments   xi

1 Stormy Weather   1
  Interview with Amanda Townley, PhD   11
2 The Great Family of Nations   17
  Interview with Paul Arthur Berkman, PhD   31
3 First the Sun   37
  Interview with Frank Niepold, MSEd   63
4 On Ice   71
  Interview with Louise T. Huffman, MSEd   89
5 The Auroras   93
  Interview with Meredith Goins, MSIS   109
6 The Shifting Earth   115
  Interview with Spencer Weart, PhD   129
7 Rockets and Satellites   133
  Interview with Rebecca Charbonneau, PhD   149
8 The Dawn of Space Science   157
  Interview with Katsia Paulavets, MS   169
9 The Sky Above   175
  Interview with James Marshall Shepherd, PhD   195

10 The Oceans Beneath   197
    Interview with Wallace J. Nichols, PhD   209
11 Aftermath   215

Notes   223

Index   233

Prologue: Half-Remembered

A few months ago, when I was first imagining what this book would be, I was having coffee with an old friend. Steve's a bit older than I am—late sixties—and he's intelligent, funny, and all-around good company. At one point in our conversation, I mentioned that I was working on a new book (this one!), and he asked what it was about.

I hesitated for a split second, fully expecting to have to explain the obscure topic to someone I felt sure would never have heard of it. After all, no one else I had told about the book had recognized what I was talking about.

I leaned back in my chair and smiled to myself. Here we go . . .

"Have you ever heard of something called IGY?" I cagily asked my friend. And then the unexpected happened. The puzzled frown I had expected never materialized. Instead, Steve looked past me for a second, as if something completely surprising had just flashed across his field of vision.

"International Geophysical Year!" he said with a curious mix of confidence and uncertainty.

I must have looked astonished, as he gave a little smile and continued: "Yeah, sometime in the late 1950s, right? It was this big scientific event." He paused for a moment, gathering the wisps of memory, while I remained silent, not wanting to interfere with his recall. "I just remember it was really important. I read about it at school, in the *Weekly Reader*."

That made me laugh in surprise. Not only had Steve retrieved a distant memory of the International Geophysical Year (IGY), this "really important scientific event" I was going to be writing about, but he had also invoked one of the highlights of my 1960s elementary school experience, the day that my teacher would hand out the current events news magazine written especially for me, Steve, and millions of other schoolkids, the *Weekly Reader* (aka *My Weekly Reader*).

As I explained to Steve what IGY was, what it meant, and why I thought it was worthwhile to write a book about it, he remembered more and more fragmentary images and ideas from that *Weekly Reader* moment. And yet his memories were not complete—far from it. To Steve, IGY was a half-remembered event he read about as a young schoolboy (and no doubt discussed in depth in class). Still, that half-remembered day when schoolboy Steve had read about a captivating event called IGY made it clear to me that the International Geophysical Year had touched the lives and inhabited the memories of many millions of people around the world, far more than I ever expected, whether they were conscious of it or not. Steve's recollection of IGY, jarred loose after decades, solidified my desire to write a book about the half-remembered International Geophysical Year.*

So what was IGY, and why did the editors of *Weekly Reader* deem it important enough to inform America's schoolkids about it? Well, it was certainly "international," as it involved the organized and co-ordinated efforts of thousands of scientists around the globe. And all those scientists specialized in what we can broadly categorize as the geophysical sciences: geology, meteorology, astrophysics, biology, oceanography, geomagnetism, seismology, and more. The "year" part proved to be somewhat off-kilter, as IGY encompassed the latter half

---

* Not in its entirety, however. The official 1965 *Report on the U.S. Program for the International Geophysical Year*, which I have used for reference, runs to over nine hundred pages. Try as I might, I simply cannot explore every facet of the U.S. IGY project contained in the report, but I do my best here to represent the totality of the event, so the reader can fully appreciate the time, energy, and spirit summoned by Matthew Maury in his pioneering attempt to unify the meteorologists of the world.

of 1957 and the entirety of 1958, due in no small part to the difficulties and delays involved in putting the first artificial satellite into earth orbit (more on that subject later).

Hugh Odishaw, the leader of the U.S. IGY program, described the event at that time as "the single most significant peaceful activity of mankind since the Renaissance and the Copernican Revolution," but it wasn't the first time such a feat had been attempted. Twice in the 1800s and then again in 1932, scientists agreed to pool their knowledge, skills, and experience to study one particular area of global scientific interest and concern.

The first scientists to see the possibilities of this approach were the world's weather experts, who recognized in the mid-1800s that the earth's atmosphere was a unitary system, fantastically huge and endlessly complicated, spanning the globe and existing in a constant state of change. Even then they understood that what affects the weather in one part of the world also affects the weather in every other corner of the world. "A storm that flooded cities in Ohio or leveled crops in France was not an isolated phenomenon," wrote *New York Times* science writer Walter Sullivan in 1961. "It was a by-product of vast forces at work within the atmosphere."[1]

There was another vast force at work here as well: unfettered capitalism. Those flooded cities in Ohio and those decimated crops in France all cost a lot of somebodies a lot of money, and that needed to change (more on that subject later). The solution to this problem took root in 1800 in France, where a network of weather observers coordinated their efforts to record local weather data that could then be standardized, shared, and compiled. Although predicting the weather was a distant fantasy in the 1800s, there was value to be derived from the publication of weather observations, analyzing what had already happened in the atmosphere in lieu of postulating what could happen.

Eventually the concept of coordinated weather data collection and analysis migrated to the United States, under the auspices of U.S. Navy commander Matthew Maury. Widely considered the father of modern oceanography despite being unable to serve at sea after being injured in a stagecoach mishap, Maury found plenty to do on land. He was instrumental in the founding of the United States Naval Observatory in 1893 and of the Navy's Hydrographic Office in 1866. During this time, Maury pioneered the science of hydrography by developing a method of mapping the bottom of the Atlantic Ocean that involved dropping cannonballs with attached lengths of twine into the sea and measuring the rate and limit of their descent.

But Maury had more to measure, and he intended to carry out his measuring on a global scale. By 1853, he had persuaded the navies of Belgium, Britain, Denmark, France, the Netherlands, Norway, Portugal, Russia, and Sweden to join the United States in using standardized forms for recording weather and oceanic events and then to share the records for the mutual benefit of these seafaring nations. As an indication of the importance of this data and of Maury's powers of persuasion, he convinced the participating navies to continue with their observations and recordings even if their ships were captured by an enemy during wartime.

"In harnessing and analyzing thousands of scientific observations from around the world, the results of [Maury's] work revolutionized our understanding of oceanography, meteorology, and marine navigation," wrote historian Tim St. Onge.[2] Regarding the success of the program, Maury enthused, "Rarely before has there been such a sublime spectacle presented to the scientific world; all nations agreeing to unite and cooperate. Though they may be enemies in all else, here they are to be friends."[3]

Maury laid the groundwork for what international cooperation in the earth sciences could look like, forever changing the texture of our lives and our relationship to our planet. But Maury's work was only the beginning. To face the global challenges that lay ahead, the world's scientific minds would need to put a much grander effort into play.

Acknowledgments

My first thank-you goes out to my literary agent Wendy Keller, who has boosted and guided my career off and on for the better part of thirty-four years. She always seems to appear just when I need her, she has an uncanny ability to make me feel as though I'm her only (and most talented) client (which I am most certainly not), and she has never, ever steered me wrong.

Old friend and clean air colleague Steve Sokolsky gets a shoutout for reaching back to his childhood to recall reading about the International Geophysical Year in an issue of *My Weekly Reader*. Steve's elementary school memories of IGY convinced me that there was a wonderful story to be told here.

Many thanks to my friend, former occasional neighbor, and ice whisperer Jim Koehler for his insights into the nitty gritty details of his work, for which he spends good chunks of every year drilling holes in the Antarctic ice so the glaciologists can do their stuff—literally, the coolest job imaginable.

I never could have told this story without the input and assistance of the International Science Council (ISC) and the American Institute of Physics (AIP). These organizations were two of the very first contacts I made when beginning my geophysics research, and they both responded to my requests with unending patience and enthusiasm. I am especially grateful to the people at AIP for allowing me to dive into their amazing library of recorded oral histories of scientists who lived IGY as it happened. AIP's foresight in preserving these memories for all time is simply astounding. Every time I started reading the transcript of one of these interviews, I felt as though I was touching history.

On behalf of every member of the human race, I wish to thank the entire staff of the World Data System (WDS), who, sixty some years after the end of IGY, still strive tirelessly to preserve and protect the data collected by those thousands of scientists around the globe who worked so hard to understand our planet and our star. I find it awe inspiring that WDS continues to this day to make that data available to scientists and thinkers everywhere.

Her many contributions to IGY may never be fully acknowledged, so I want to offer my thanks here to Abigail Van Allen, wife of James, who made dinner that night in 1950 for a gathering of preeminent physicists. While Abigail toiled in the kitchen, the scientists were cooking up the idea of the most audacious scientific undertaking of all time, and, by all accounts, it was the dessert—Abigail's "fantastic" homemade chocolate layer cake—that brought the group to a consensus: 1957 would be the year that science changed everything.

CHAPTER 1

Stormy Weather

Once upon a time, an American city on the Eastern Seaboard found itself besieged by multiple extreme weather events within a seventy-two-hour period. First was a hurricane, accompanied by multiple tornadoes; then a riptide warning; then flash flooding; then a "blue moon" with accompanying high tide; and last but not least, a full-scale geomagnetic storm hurled our way from a pitiless solar eruption. Damage was in the millions, there were four fatalities and many injuries, and thousands of people lost power and water.

The hurricane—Idalia—narrowly missed the center of the city—Savannah, Georgia—which got off easy compared with dozens of cities and towns in the region that were slammed head on by the hurricane as it wore down, surprisingly quickly, into a tropical storm. They say we got lucky; it could have been far worse.

I was in Savannah that day, and although I grew up in tornado alley in the upper Midwest of the United States and am usually unfazed—even excited—by funnel clouds, this was my first hurricane, and I was appropriately nervous. Maybe that's because at the peak of the turbulence the TV weatherman said, "If you're tempted to step outside to see if you can see the storm funnel, don't, because it's already on top of you and you won't possibly get back inside in time." No need to tell me twice, although he did. As I sheltered in the center of my dwelling, under a door frame and away from the windows, I could feel Idalia's approach unexpectedly *inside* my body. As the atmospheric pressure crashed, so did my internal organs. I felt compressed inside—not crushed, exactly, but strangely diminished, almost shrinking. For one weird instant I pictured my inner organs as a big, shiny, heart-shaped mylar balloon that was collapsing as all the air was sucked out of it. A moment later the compression sensation went away, to be replaced by what my siblings and I used to call an "ice cream headache" or "brain freeze" when we were kids, a dull, heavy pain that centered on our sinuses and spread between our shoulder blades when we would eat ice cream too greedily, which was every time we ate ice cream.

After ten minutes the brain freeze faded away, and soon the storm passed. The hurricane dwindled to a tropical storm somewhere over South Carolina and then turned out over the ocean. The other weather events were anticlimactic to me. I wasn't tempted to go swimming in the raging Atlantic that day, so the riptide warning didn't apply. Neither did the blue moon and the high tide. The tornadoes all dissipated, as they do. The flash flood warning could have been a concern, as I was only a mile or so from the coast and the rain was relentless for much of the day, but the floodwaters failed to materialize—at least in my neighborhood.

However, the geomagnetic storm is actually at its peak as I write these words, three days after Idalia passed over us. Caused by two near-simultaneous coronal mass ejections (CMEs) from the sun, the storm is expected to hit the earth's magnetic field within the next twenty-four hours, potentially interfering with power grids and radio transmissions, and triggering a rare display of the northern lights (aurora borealis) for lucky skywatchers to our north.

That's a fairly detailed overview of the last seventy-two hours of my personal weather news, made possible by a formidable army of weather satellites, TV meteorologists, ground-based radar systems, amateur storm watchers, both private and dot-gov websites, and a big helping of just plain choosing the right moment to peek out the front window to see what's going on out there to possibly witness the transit of a flying car or an uprooted tree. For all those things I am grateful, and even in awe.

Now, imagine that same scenario back in the early 1800s. At that time, my only source of information about the death and destruction approaching me from the sky would have consisted of a compressed feeling in my upper torso, brain freeze, and me peeking out the front window (if I had one). That's it.

Which all goes to illustrate that we humans have—and always have had—a rocky relationship with the world around us. And not just around us but *in* us as well, as my compressed internal organs and brain freeze can attest.

This situation has been troubling people for centuries. At a scientific and medical conference held in Graz, Austria, in 1875, Karl Weyprecht, a lieutenant in the Austrian navy and an avid Arctic explorer, presented the concept of the First Polar Year to his fellow scientists. Weyprecht's plan was to establish research bases around the Arctic and Antarctic circles, to make continuous observations and recordings of the weather, the movement of polar ice, the earth's magnetic field, and the ever-mysterious northern lights. Up until this point these types of observations would have been conducted by lone scientists on a local, isolated level; Weyprecht's innovation was to synchronize observations by numerous scientists around the globe, generating simultaneous data from every base at exactly the same time.

Weyprecht's plan, brilliant as it was, was sidelined by a war between Russia and Turkey that essentially brought international scientific research to a screeching halt. It took several years, until 1880, for his ambitious plan to come to fruition at an International Polar Conference. There, representatives from eight countries agreed to establish fourteen polar stations that would make coordinated weather observations for one year, beginning in August 1882. Dubbed the First Polar Year by its optimistic organizers, who clearly expected there to be more than one, the event was scarred by the deaths of many participants, including seventeen members of the U.S. scientific team based at Canada's Ellesmere Island. Stranded on the icepack for several years and facing starvation, amputations, and death by exposure, the surviving Americans nonetheless continued their observations until the bitter end, going so far as to set irreplaceable scientific instruments and duplicate records of their observations adrift in separate boats in an attempt to safeguard their work.

Fifty years later, a proposal for a Second Polar Year stated that the sequel would "be of practical application to problems connected with terrestrial magnetism, marine and aerial navigation, wireless telegraphy, and weather-forecasting."[1] This event encountered its own misfortunes, however, falling victim to global economic and technical challenges. First was the prospect of a celestial missed opportunity, as the event's planned 1932–1933 timeframe would completely miss the peak of the eleven-year cycle of sunspot activity. This was deemed a critical issue, as sunspots—bundles of magnetic energy that appear regularly on the sun's surface—were known to cause magnetic storms in the earth's atmosphere. Second, world governments were reluctant to fund such an expansive and expensive scientific research project in the midst of the Great Depression.

Still, the backers of the Second Polar Year managed to recruit scientists from forty-four nations and to use the new technology of radio to advance the cause of weather research. But although radio provided scientists with a valuable new tool for conducting and reporting research, it also presented them with a conundrum. Radio waves travel in a straight line, but the surface of the earth is curved. How is it possible, then, that one can receive a radio signal sent from the other side of the globe? The phenomenon suggested that radio waves were being redirected back to the earth by a hitherto unknown "mirroring" layer in the upper atmosphere and could thus reflect their way around the curvature of the earth. "It was becoming clear that, to understand weather in the lower atmosphere, one should learn what goes on in the upper air," Sullivan wrote.[2]

Despite the mixed results from the first two Polar Years, the idea behind these scientific milestones did not lose support, even if it lost some momentum. That momentum got an unexpected boost at a social gathering in April 1950, at which a group of scientists were discussing the challenges of upper-atmosphere research—including ongoing study of the northern lights, the earth's magnetic field, and that "mirroring" layer that affected radio signals—and contemplating ways to better study these mysterious phenomena. These scientists had all been involved in the Second Polar Year and, at the suggestion of physicist Lloyd V. Berkner, were discussing what was starting to become a sort of scientific tradition. Instead of the fifty-year gap between the First and Second Polar Years, however, Berkner proposed a Third Polar Year that would occur only twenty-five years after the second. Not only would this timing align with the peak of the eleven-year sunspot cycle, but it would (in theory, at least) allow these men of science to complete the work they had begun as part of the 1932 program.

One of the scientists who was present that April evening was Sydney Chapman,* a celebrated British geophysicist who was in the midst of moving to the United States. Chapman was respected and admired in scientific circles for creating the first model of the ionospheric layers (created by charged particles emitted from the sun; we'll find out more about them later) enveloping and acting on the earth, and thus he was the perfect audience for Berkner's Third Polar Year pitch.

Chapman was quickly sold on the idea, and there was broad agreement among the scientists that several crucial areas of research were being held back by the paucity of available data. It wasn't that the data wasn't there; the problem was one of access. New scientific devices and techniques were producing more data in more fields than any one scientist or laboratory could keep up with or make use of. Furthermore, political and geographic differences often made data sharing problematic, if not impossible.

It was agreed that Chapman and Berkner should pitch the idea of a Third Polar Year at an upcoming scientific conference in Brussels, Belgium. Response to that proposal was "modest" at first, but rather than downsize the concept, Chapman and Berkner doubled down and

---

* Chapman relied on forty-year-old data from the First Polar Year to hypothesize the presence of currents in the upper atmosphere, proving that there was lasting value in the concept of worldwide data sharing between scientists.

expanded the scope of their idea. With growing enthusiasm and support from the membership of influential nongovernmental scientific unions, talk soon centered on expanding the scope of the event to be a study of the whole earth, its oceans, its atmosphere, and the sun, not just the polar regions.

In a July 1955 memorandum from the National Academy of Sciences, the newly defined global mission of the event is spelled out clearly: "Our environment, particularly the atmosphere and the oceans, affects the daily lives of all individuals, the transaction of commerce and industry, the safe conduct of land, sea, and air travel and transportation, and the range and reliability of all radio communication and navigation systems. One of the controlling fields in this research program is solar activity, for the sun dominates activities on our planet and is the major source of energy for the earth and for all life."[3]

Despite this bold mission statement, as more scientists identified new fields that could benefit from the Third Polar Year, the more apparent it became how shockingly little we humans actually knew about the world we inhabited. This situation simply would not stand.

With growing confidence in the endeavor, slow but steady increases of financial support, and rising public interest, Chapman rechristened the project the International Geophysical Year, or IGY, and in July 1953 he and Berkner were selected by members of several prominent scientific societies to administer the daunting event. "The greatness and simplicity of the basic purpose of the International Geophysical Year—the common study of our planet by all nations for the benefit of all—made a strong appeal," Chapman wrote.[4]

"The International Geophysical year was conceived as the greatest attempt men have yet made to band together to examine, without passion or undue rivalry, their environment, their home and ultimate resource, the earth," echoed geophysicist J. Tuzo "Jock" Wilson.[5]

But where to start? One of the first decisions to be made by Chapman, Berkner, and the nine-member committee they led was to establish the timespan of the event. "Berkner gave good reasons for the renewal of the International Polar Year after 25 years," Chapman wrote. "The Second Polar Year had fallen at a time of sunspot minimum, whereas 1957–58 was expected to be at or near sunspot maximum, when many influences of storms on the sun affect the earth." In a happy coincidence, Chapman observed that there would also be a respectable number of solar and lunar eclipses throughout the 1957–1958 period.

A scientific panel with the awkward acronym CSAGI (for Comité Spécial de l'Année Géophysique Internationale, more simply known as the Special Committee) was formed, and it ruled that the IGY would officially last eighteen months, from July 1, 1957, to December 31, 1958. As Sullivan relates, the reason for creating an eighteen-month "year" was that "twelve months was not long enough to obtain an adequate sampling of data."[6]

Under the Special Committee's guidance, fourteen scientists, referred to as "Reporters," were chosen to oversee and coordinate research and exploration efforts in the following fourteen areas of inquiry:

1. world days and communications
2. rockets and satellites
3. meteorology
4. geomagnetism
5. aurora and airglow
6. ionosphere
7. solar activity
8. cosmic rays
9. longitudes and latitudes
10. glaciology
11. oceanography
12. seismology
13. gravity measurements
14. nuclear radiation

This was in stark contrast to the scope of the Second Polar Year twenty-five years earlier, when the areas of research and experimentation were limited to only five: weather, the aurora, the earth's magnetic changes, the ionosphere, and cosmic rays. The tools and techniques to go even further than these areas simply hadn't been created yet, but that situation would change dramatically by the 1950s. Chapman illustrated the progress of science linking IGY with the two earlier Polar Years, pointing out that during the First Polar Year observations were limited to ground level. Fifty years later, during the Second Polar Year, scientists could observe the earth from an altitude of six miles using balloons. At the time of IGY, twenty-five years later, weather observations could be made by rockets launched one hundred or more miles above the earth. Indeed, one sage noted that if the earth was the size of a desktop globe, the regions of the atmosphere that we had explored by the mid-1950s would be no deeper than "a heavy coat of paint."

Once the timetable and topics of study were agreed on, the next step was to determine how to use, share, and safeguard the immense amounts of data that would be generated over the eighteen months of IGY. At the conclusion of World War II, it was discovered that many European records from the Second Polar Year had been lost or destroyed, and CSAGI was determined to avoid a similar fate for IGY records. In order to protect the future legacy of the scientists of IGY, all observations were to be stored at a network of interconnected world data centers (WDCs), each managed by a coordinator who would oversee the preservation and distribution of data collected at the center.

Any and all scientists and scientific organizations, whether they were IGY members or not, would be able to acquire research data from any WDC, for only the cost of reproduction and postage. The United States and the USSR created duplicate data centers that stored all data from all IGY projects and experiments, while a multitude of other member nations maintained smaller WDCs that housed the data pertaining to one or more IGY fields of research. Thus, the data from any one field of research would be safeguarded in at least two WDCs, likely three or more.

Sullivan described IGY as a sort of "scientific club," and there was some truth to that. For a research program to be included as a part of IGY, it had to meet certain criteria, and only specific types of scientific problems would qualify:

a) Problems requiring concurrent synoptic observations at many points involving cooperative observations by many stations.
b) Problems of branches of the geophysical sciences whose solutions would be aided by the availability of synoptic or other concentrated work during the IGY in other geophysical sciences.
c) Observation of all major geophysical phenomena in relatively inaccessible regions of the earth that can be occupied during the IGY because of the extraordinary effort during that interval, in order to augment our basic knowledge of the earth and the solar and other influences acting on it.
d) Epochal observations of slowly varying terrestrial phenomena, to establish basic information for subsequent comparison in later epochs.

Priority would be given to research projects in category *a*, as they involved coordinated observations from multiple posts and thus could create the most valuable data. Projects in category *d* were important for their inclusion of past research data and their focus on the preservation of data for future researchers.

There were several areas of inquiry that meshed with these requirements, but the ones that captured the most attention and imagination in both the scientific and the public spheres were:

- Antarctica
- the Arctic
- outer space
- the equator
- three pole-to-pole meridians (one through Europe and Africa, one through the Americas, and one through Asia and Australia)

And so the wheels were set in motion, taking science to a place where it had rarely ventured. "The question I am most frequently asked in connection with the IGY is whether there was genuine exchange of information among nations, and especially between East and West," wrote Wilson, looking back on the experience. "There *was* a genuine exchange, and it was possible because the men and women concerned had been accustomed for years to meet in a friendly atmosphere to discuss new information, ideas, and discoveries. The International Geophysical Year was but a vaster, more comprehensive version of these regular sessions, conducted in the same atmosphere of scientific inquiry." "It is a great deal simpler to work out a plan for charting the world's oceans," concluded Wilson, the official chronicler of IGY, "than it is to discuss freedom, wealth, politics, or religion."[7]

Wilson wasn't wrong in that, but even from his bird's-eye view he couldn't anticipate the immense challenges that lay ahead, both above and below the world's oceans.

Interview with Amanda Townley, PhD

Executive Director of the National Center for Science Education

Nearly seventy years since the IGY's final curtain call, we are still left with puzzling questions and monumental concerns. Under the auspices of the International Arctic Science Committee (IASC) and the Scientific Committee on Antarctic Research (SCAR), a new International Polar Year (the fifth) is being planned for 2032–2033 "to close outstanding major knowledge gaps" and address issues of climate change, but will that be enough time to move the needle as far as it needs to be moved?

With this question in mind, I conducted a series of interviews with scientists, researchers, and educators whose work has a direct impact on our understanding of climate change. I chose my interview subjects based not on their fame or visibility in the scientific world or public arena but because their work interested and intrigued me and made me think more broadly about the message I hoped to send with this book.

Although the wording may change from interview to interview, my primary question to each of these experts was the same: If thousands of scientists from around the world could set aside their biases and egos and join forces to address some of the most challenging scientific issues of 1957 (and capture the world's imagination as they did so—no small feat), could the same thing be done today?

Some of "my" experts expressed confidence that it could be done, some even pointing out that this is already happening in various scientific spaces around the world. Some were not so sure it could be done but held out hope that in time that could change. Along the way, several of them added fascinating and unexpected new perspectives to climate science, such as

- respect for and utilization of indigenous knowledge,
- inclusion of social sciences as an aspect of climate research, and
- the power of "science diplomacy" in the promotion of international cooperation.

And in the end, all believed that IGY was of tremendous benefit to humankind and was, in fact, the year that science changed everything.

Mark O'Connell: What are the objectives of the National Center for Science Education?

Amanda Townley: The focus of our organization is ensuring that teachers have the ability as well as the confidence and skill set to accurately teach topics that the public often sees as controversial. So things like climate change, climate science, evolution. And then also we target the nature of science, because very often, it's misunderstandings of what science is, and is not, that are actually used as arguments for anti-science movements.

So we operate in three different areas: we operate to investigate science education. If you've ever heard of the *Making the Grade* report that looks at standards for climate change, and how different states are preparing young people to make decisions about climate and things like that, that's our organization.

We also have a catalyzing action wing that monitors any legislation across the country that aims at inaccurately representing topics like climate science, in their standards for teachers.

And then, of course, we have a direct teacher wing that actually provides a snapshot curriculum that can be integrated into any curriculum approach in any school in the country, to help teachers better teach climate education.

Our message relative to climate change is, it is real. It is bad, but there is hope. You know, people don't act; they suffer from apathy, and they think there's nothing we can do. It's out of our hands. But in reality that's not true. There's plenty of hopeful moments out there; we just have to actually look for them.

M.O.: When you take part in an event with school kids, do you ever see any emerging or potential new Greta Thunbergs among these kids?

A.T.: You know, our kids are pretty amazing. And very often, they're more open to learning about climate change, they're more hungry for action and learning than the older generations are.

So we had a number of students who did student presentations during the event, and they were talking about the different civic organizations related to climate that they were engaged in, and their independent research that they were engaged in as high school students. And it really instills a great deal of hope, that instead of having one Greta Thunberg, we could have this whole generation of people who are thinking and acting on that.

M.O.: Do you see any signs that scientists would resist coming together to work on addressing climate change? If somebody proposed this idea now of everybody dropping their work and coming together? Do you think that could work? Am I just being naive when I dream about something like this happening?

A.T.: No, I absolutely don't think it's naive. Scientists are way more cooperative than what we're giving them credit for, in terms of working groups and things like that. Very often, the barriers to scientists working together are barriers that exist because of national boundaries, political arguments that are well outside of the work that we do. Science in and of itself is apolitical, it's more about knowing where the evidence leads. And if there were to be the ability to call world scientists together and say, hey, you know, we have the funding for this, you know, all of these external barriers to your work are gone, will you come together? The answer would be a resounding yes. Because to a great extent, they're already doing that. They're sounding the alarms together on these climate issues, specifically.

M.O.: In what ways are you seeing science being activated to address climate change?

A.T.: Science has always been a social endeavor. The idea that science is done by one person sitting in a lab somewhere and doesn't involve communication with others, and data from so many places [isn't accurate], especially in the modern world where we're all so connected. But there are a lot of barriers. There are countries who aren't allowed to share things, they're not allowed to share their data. There are different institutions and implements that are in place that prevent the complete widespread sharing of all the things, and if those things were to be set aside to the extreme that they were for IGY it can be tremendous.

But I also think our bigger challenge is the billion-dollar misinformation industry around climate change. It's literally a billion-dollar industry, with all of these different entities and special interest groups, doing everything that they can, not just politically, not just in the news, but even in things like school curriculum. One of the things that we're fighting in specific states is curriculum with cartoonized versions of fossil fuel advocates who are saying that climate change is good for the planet.

M.O.: I would imagine that most, if not all, of the kids would see through that immediately, would know that they're being fed a lot of bull. Or, again, am I being naive?

A.T.: Well, we would hope, but when that's implemented as curriculum. . . . For example, Florida just approved the use of PragerU videos in classrooms that literally say that climate change is good for the world and that we shouldn't do anything about it. So states are adopting this as their curriculum. And, you know, if you're in a place where your parents don't know that much about climate change, or they're getting misinformation from their sources, we can't assume that misconceptions are just going to magically fix themselves, because they don't; if we don't target misconceptions head on, they persist all the way through adulthood.

And so you would be surprised, they really don't. As far as Climate of HOPE, over four hundred teachers were at that event. Wow. And so being able to help support them and getting accurate information out to their students, and helping them build confidence to say, hey, you know, this source that you keep seeing on the news, this thing you keep seeing in the media, you know, that saying humans did not do this? Well, here's what the scientists say. And when your students are bombarded in curriculum that's inaccurate, if they're bombarded with constant messages on social media and other spaces where they're getting their information from parents and trusted adults when they're younger, and then trusted friends as they get older. And so that's probably the most dangerous assumption we can have is that people are just going to magically figure it out.

M.O.: Do you work only with K–12 students? Or do you do extend your work into higher education?

A.T.: The National Center for Science Education is focused on K–12, but we also engage with preservice teachers and undergraduate education. Most of our work is not with students, it's with their teachers, because every teacher will impact, depending on their grade level, hundreds to thousands or more students in the course of a career. Being able to work with teachers, and ensure that what they're teaching is accurate, is a lot more effective than, say, a one-off seminar or things like that with a group of students.

M.O.: Is there much interest or understanding among students of science history? It's almost as if there's this barrier, pre-1957 and post-1957, where attitudes and understanding just seemed to shift. And I think in large part because of IGY, do you see any interest or aggressive activity in studying scientific history?

A.T: The way science is traditionally taught is from a very historical perspective, but not really to that level because students have one earth and space science, for example—in Georgia earth and space sciences, sixth grade, that's all they get. So unless they sign up for an environmental science course, or an earth science course, in secondary education as an elective, they're taking mostly life science and physical sciences. And so the history of science, philosophy of science, nature of science, are generally not a strong inclusion in K–12 education.

We find that in higher education, even in the sciences, a lot of times students don't get a philosophy or history or nature of science class. And so they learn a lot about what we learned, but they don't learn the historical context of when that happened, and how that came about. Okay, which is a shame, because you miss a lot of those frameworks, like what you're talking about with IGY, where it was literally this massive melding of the minds.

And so the context of that was just so world changing, and massive, because it allowed this massive sharing across boards that otherwise would have never been able to (work) together. Whether it's the 1930s or 1950s, each of these different event horizons was in a lot of ways like the equivalent of a modern internet for those days, because now we communicate so much more freely, even though there's still barriers and boundaries. Imagine, again, you have no internet, you have telephones, you can call within a certain limit, but a lot of these countries weren't communicating with other countries out of fear.

M.O.: To end this conversation on a high note, what can you tell me about Insta hope?

A.T.: Teachers think a lot about the fact that their students use this wide range of media to get their information. It's a very different world from the world in which many teachers grew up. Now it's short clips here, short clips there, TikTok, social media; all of these different places where our kids get their information and engage with each other look very different from what it used to be. So I specifically point out some incredible Instagrammers that are young people who are doing amazing things. One of my favorites is the Garbage Queen (https://www.goodgoodgood.co/articles/alaina-wood-garbage-queen).

She's an amazing young woman; she does climate hope news every week, with an audience of 194,000 followers. Yeah, it's worth a look, she highlights things that are going right. Insta hope is just a snapshot of some of these different Instagram accounts of young people who are using these platforms to reach out to other young people to educate, to inspire, to help drive actions.

They have a voice now, they have this ability to get a message out there. And some of these people are choosing these wonderful messages of hope that are super inspiring.

CHAPTER 2

The Great Family of Nations

Lloyd Berkner and Sydney Chapman, having had greatness thrust upon them, realized they had a monumental challenge ahead of them, with a mountain of monstrous questions to answer and issues to discuss. "Too many people were aware that hitherto observations had been made haphazard, with instruments which had not been standardized one with another, the results written up in a form which made comparison with similar observations unduly difficult," observed science writer Ronald Fraser, putting Berkner's and Chapman's task in stark relief: "How can the scientists of the world, the people who will have to *do* the job, be most directly reached? Who should make the first outline of the programme, and who should execute it? When should the attack begin?"[1]

In Fraser's opinion, Berkner and Chapman were the best people to take on the challenge posed by IGY, and they would quickly prove to be excellent partners. "Chapman is the academic, integrating brain," Fraser observed, adding that "swimming and cycling have kept him remarkably fit, and he walks today with the gait of a young man."

Fraser described Berkner as a man of action. "He was radio-officer with the expedition of Admiral Byrd to the Antarctic in 1928–30" and was instrumental in the U.S. Navy's adoption of radar technology. "Both were agreed," Fraser wrote, "that the vessel which should carry the flag of the effort should be the International Council of Scientific Unions."

The council, generally referred to as ICSU (rhymes with "Fix You"), was not controlled by any government body, and it consisted of a group of international scientific unions and different counties' scientific academies, forty in all. The unions were largely self-funded groups made up of scientists working together on very limited budgets to advance the study of their particular field of research, while the national academies were generally well funded by their government's treasuries and supported work done in many branches of science. With a rough structure in place and Berkner and Chapman fully in charge, the special IGY committee met in Brussels in 1952 and officially lit the fuse for the now-inevitable event. Brussels became the home base for IGY as the growing number of participating unions and academies worked together "to link the programmes put forward by the [then] fifty-four nations with the general world-wide programme which would benefit them all."[2]

"Each participating country is planning and developing its own program, and the results obtained will be made available to the scientists of the world," announced a press release issued on July 19, 1955, by the National Science Foundation. "The National Academy of Sciences–National Research Council, which represents the interests of United States scientists in the International Council of Scientific Unions, is responsible for development of the scientific program to be undertaken by this country during the International Geophysical Year. Special Federal appropriations being made in support of United States participation in the International Geophysical Year are being administered by the National Science Foundation which is coordinating government interests in the over-all program."

From the beginning, IGY promoted interdisciplinary research, as seen in a report to the National Science Foundation:

> Scientists in many research institutions engaged in studying IGY data, attempting to apply this data to problems in three general areas:

a. The earth itself as a structure
b. Atmospheric and oceanic circulation and heat and water budget of the earth
c. Upper atmosphere physics and solar-terrestrial relationships

As part of this general program of study of IGY data, many distinguished scientists from other countries visited their colleagues in the United States, discussing problems of interpretation and investigating fruitful avenues of approach to those great geophysical problems which scientists the world over are attempting to understand.[3]

As mentioned in chapter 1, the ICSU created a governing body for IGY, dubbed CSAGI. With astonishing speed, CSAGI turned the nebulous into solidity. IGY would be funded by all participating countries to the tune of $250 million, and all countries of the earth, whether they participated in IGY or not, would receive "very practical dividends in return," according to Hugh L. Dryden, a trustee of the National Geographic Society:

They will enjoy improved weather forecasts and radio communication. They will benefit from greater knowledge of the upper air and nearby space in which airplanes, satellites, and eventually space ships will travel. Moreover, there is the possibility of far greater unsuspected discoveries of who-can-guess what value to man.

The International Geophysical Year will make a significant contribution to man's increasing search for clearer understanding of his surroundings. It will see the first faltering steps toward man's exploration of outer space. It will stir the imaginations of countless boys and girls to the wonders and opportunities on the road ahead toward the far horizons of space.[4]

In October 1952, CSAGI sent out the first invitations to participating nations to submit their plans for IGY. Soon afterward, newly formed working groups began to meet in various IGY countries to brainstorm their scientific goals. Oceanography was the topic at a

January 1957 meeting in Göteborg, Sweden, while nuclear radiation was discussed at Utrecht, Netherlands, the same month. Geomagnetism was the subject of a meeting in Copenhagen, Denmark, a few months after that, and still later a conference on rockets was held in Washington, D.C.

As the concept of IGY caught hold and became more real to the world's scientists, there was an almost giddy sense of excitement in the air.

Astrophysicist Marcel Nicolet, PhD, secretary general for the committee, announced that in these early planning meetings "the atmosphere in CSAGI circles was characterized by great enthusiasm, and almost a sense of euphoria."[5] In a February 1956 essay for the National Geographic Society, Hugh L. Dryden, director of the National Advisory Committee for Aeronautics and a National Geographic Society trustee, described IGY as a "great symphony of science" and rhapsodized that "the scientists will work as a well-conducted orchestra in harmony."[6]

Dryden's letter, published nearly a year and a half before the IGY start date, offered a sneak preview of how IGY would work and what might be learned from the project. First, he offered readers a basic explanation of just what geophysics entails. "The scientific study of the motions, heat, light, and electrical properties of our earth, its continents, oceans, and surrounding atmosphere is the science of geophysics," he wrote. "The earth is too big to bring into a laboratory, and man as yet has little control over storms and earthquakes. The earth scientist can only observe the experiments which Nature makes, and he himself can only see what happens at one place at any one time."[7]

"The CSAGI asked that each nation organize a national committee to advise how its nation could contribute to the achievement of specified IGY objectives and to plan and supervise its national contributions," wrote Berkner. "Through integration of the national plans, and negotiation and appeals to fill the gaps, the CSAGI forged the world-wide program of simultaneous geophysical observation now specified in the *Annals of the IGY*. Several points about this method of organization are worthy of notice:

1) *It is successful in catalyzing extensive research.* This specification of particular objectives tends to inspire the desire for participation by every national group, since it satisfied a sense of national aspiration.

2) *The organization of the research is through national machinery.* Consequently, the governments of the world respond favorably to requests by their own national committees for support of specific planned [*sic*] and endorsed by the world's leaders in science but carried out by the individual nations.

3) *Research can be stimulated that would not otherwise be done.* As examples of the power of cooperative planning, the scientific exploration of the Antarctic continent and instrumented earth satellite stand out. Moreover, the value of the research of one nation is augmented by the corresponding and supplementary planned researches of other nations.

4) *The organizational machinery of the Special Committee is financed from international funds,* in which UNESCO (United Nations Educational, Scientific and Cultural Organization) has properly played a leading role. The amount of such administrative funds is small (a few hundred thousand dollars), compared to the huge research funds of hundreds of millions of dollars that are catalyzed by the international stimulus.

5) *The Special Committee is entirely nonpolitical.* It states overall objectives, plans, and requirements, but makes no recommendations that would involve one nation working in the territory of another. Instead, it encourages bilateral or multilateral negotiations where cooperation between specific nations would be advantageous. It adheres to the principle of universality by permitting any bona fide scientific group representing the science of an area to adhere to the plan.

6) *The method captures the imaginations of the world's best research scientists.* The working together of scientists of many nations toward a common objective inspires enthusiasm for otherwise unattainable goals. It is the stuff of which the Renaissance was made."[8]

Capturing the imaginations of the world's best research scientists is certainly a worthwhile, and no doubt challenging, ambition. But what Berkner didn't mention and perhaps failed to realize was that IGY would also capture the collective imaginations of many nonscientists who were curious about how the world works, and in an unexpected way, making great strides in straightening out and ultimately unifying the human race.

Public interest in IGY was growing before even a single balloon was launched, and the U.S. Congress made sure that the National Academy of Sciences would devote adequate resources to boost its visibility. First, the academy launched the publication of a monthly news journal, the *IGY Bulletin*, to be distributed to schools across America (the *Weekly Reader* came much later). To complement the *Bulletin*, the U.S. Committee for IGY produced a series of thirteen thirty-minute color films, one on each of the main areas of scientific study. The *Planet Earth* films, paid for by the generosity of the Ford Foundation, were likewise distributed to primary and secondary schools across the country and also aired on public and commercial television. The films were ultimately dubbed into Spanish, Portuguese, and Arabic and were shown on television stations in counties where these languages were traditionally spoken.

The Ford Foundation also paid for the creation of a striking six-piece poster set perfect for classroom bulletin boards, depicting the six major areas of IGY activity: the poles, oceanography, meteorology, the sun, the earth, and space. The posters were accompanied by a forty-three-page booklet for students titled *Planet Earth: The Mystery with 100,000 Clues* and a handy resource guide for the teacher. They illustrated the six mysterious realms of IGY with vivid colors, abstract designs, and, most interesting, subtle depictions of classical deities and imaginary beings controlling the actions of each realm. The "Oceans" poster, for example, pairs an image of Poseidon shepherding a ship across the waves accompanied by a quote from Lord Byron, to wit: "Icing the pole or in the torrid clime, / Dark—heaving—boundless—endless and sublime." The poster devoted to "Space" is dominated by an image of a hand appearing from the bottom edge of the poster, reaching up and seemingly trying to touch the stars above. "Ah, but a man's reach should exceed his grasp, / Or what's a heaven for?" reads a quote from Elizabeth Barrett Browning, adding a longing layer of wonder to the image. The posters, simply put, are far more captivating, far more wondrous than what you might expect to see on a bulletin board in an elementary school classroom. One can only wonder how many young minds were enthralled by the imagery and were gently steered into a life of science as a result.

"It certainly got publicity," recalled Arctic oceanographer Kenneth Hunkins of the efforts to promote IGY as a global imperative. In his 1998 oral history interview conducted for the American Institute of Physics by Mike Sfraga, Hunkins commented on the crossover appeal of the event: "People that were at all interested in science, even though they weren't really in it professionally, would be aware of it I think . . . it did get a lot of attention. It was dramatic with these snow cat parties crawling across the frozen continent.* And I suppose the National Science Foundation, somehow, anyway in later years, of course, they were very involved. They must have provided funding for the journalists to go down there and then write stories, and they've done that ever since. And they've got a lot of publicity over the years."

Hunkins's work on Arctic ice research is worth noting, if only for its wow factor. "His most well-known work emerged from a series of drifting ice camps in the Arctic," reported the Columbia Climate School, The Earth Institute, in 2014. "In 1957, he helped staff Ice Station Alpha, where the U.S. Air Force landed Hunkins and some 20 other scientific and military men by plane on the sea ice some 500 miles north of Alaska. Just below them was 10,000 feet of water."[9] Over the eighteen-month duration of IGY, the half-mile-wide slab of ice on which the base rested drifted some two thousand miles.

Such ice camps are rare today, now that satellites keep their eyes on the earth. Just the same, in 2007 Hunkins shared some thoughts about

---

* In this context, a "snow cat party" refers to a convoy of tracked snow vehicles on a traverse across the polar ice, not a group of celebratory felines.

drifting with the ice: "That's still one of the best ways to understand ice—to drift on it. You don't get that intimacy with ice unless you're living on it. . . . Of course, today, there's a lot less of it."[10]

"The Antarctic got the publicity," Hunkins mused. "I didn't see too many journalists in the Arctic although Walter Sullivan at the *New York Times* was always a faithful chronicler of activities . . . in the Arctic. And our work was mentioned in his one book he wrote about the IGY. So there was publicity.

"But I would say probably if you had to pinpoint one area, probably the Antarctic was one that got the most attention because you know it's interesting thinking of the IGY, to me. . . . [M]y idea of it was it was originally the concept of a few upper atmosphere physicists. But there was a Belgian, Nicolet, and a few people like that, Europeans, had the idea that we really had to have a global type of a year, and they picked '57, '58 as being one of the multiples of the International Polar Years. I think they'd come twenty-five years since but it was '32 or something like that. And then before that had been fifty years."[11]

"So they said, well we'll speed it up. It will be after twenty-five years," Hunkins said. "And that's why we had the IGY. And I think they thought of it as an upper atmosphere experiment. And of course the University of Alaska Geophysical Institute was very involved in that with auroral observations and so . . . they needed to get a lot of stations in Siberia and in Alaska and upper northern Canada and Greenland so that they could get global pictures of the aurora."[12]

"At that time the rockets were, from World War II were just becoming available," he continued. "I think it was just the right time after World War II. Scientific research had built up enough and when this concept of an international year began to develop, I think a lot of other people, they just joined in spontaneously. They were able to find programs within the U.S. government especially who were willing to fund them. And so, I think that the land-based ones and the ones that were looking below the surface of the earth were not what the very early founders had envisioned, but that just developed spontaneously so that it became a global experiment in every aspect of geophysics, geochemistry. So, I thought that was always interesting how it just spread. Because the upper atmosphere is very interesting, but probably the other parts of it became more interesting, the Antarctic, snow cat parties, and things like that lent themselves to journalism even better."[13]

"The outstanding feature of the IGY plan has been its completely decentralized operations with its provision for intimate and voluntary international collaboration," Berkner wrote, as if to back up Hunkins. "Thus, the plan has been predicated on the fundamental honesty of educated human beings who have come to realize that their joint welfare depends upon unselfish contribution to the advantage of all.

"Among the great family of nations, with only one exception, nothing has permitted political considerations to stand in the way of full contribution of the IGY to man's study of earth."[14]

I'll identify that notable "one exception" Berkner mentioned soon, although you may have already guessed the name of the miscreant nation.

Program planners soon found themselves facing a perplexing issue—namely, the simple fact that it's very hard to create a workable schedule of observations and measurements when it's so difficult to keep up with the earth's ever-shifting states. The phenomena to be studied by IGY scientists could be divided into two categories: forecasted, expected events, such as the position of the sun in the sky at any given interval, the moment of solar or lunar eclipses, or the precise time of the high tide; and random, unpredictable events, such as earthquakes, volcanoes, and solar flares. In a very real sense, many of the scientists would have to be light sleepers, ready to race to their telescopes or get readings from their sensors at any moment around the clock.

To minimize this issue, the IGY planners came up with the simple idea of declaring certain days of every month primary Regular World Days (RWDs) and certain periods World Meteorological Intervals (WMIs). During these special time periods, certain focused observations and measurements—instrumented rocket launches, for example—were to be conducted simultaneously by scientists around the world, thus fulfilling the promise of IGY in a tangible way. The official IGY calendar designates two days of every month as RWDs. Those days are joined by additional monthly RWDs occurring on the night of a new moon or when "unusual meteoric activity" occurs. Every quarter during a WMI, intense weather observations were to be made for ten-day spans.

The scientists participating in IGY represented a desire to synchronize the observations of all scientists around the world, gathering the same data from many different locations on the earth at exactly the same time. "The planning and execution of the IGY was marked by a most co-operative and harmonious spirit among the scientists of the 67 nations associated with it," Chapman wrote. "Their common interest in its subject and purposes made it possible for them to work together despite differences of race, creed, or political organization."[15]

"The International Geophysical Year marks the beginning of a new era in science," enthused Alexander Marshack in his IGY tome *The World in Space*. "[It is] an expression not only of man's knowledge and daring but also of his relative *ignorance* of the earth he lives on and of space above him."[16]

And about that exception to international cooperation and unity mentioned earlier, the countries in question were the People's Republic of China and its political rival Taiwan (formerly known as Formosa). Both nations were primed to take part in IGY, but both objected to the inclusion of their rival. The ire rose to the level of each country refusing to participate in IGY if the other was allowed to join.

Of course, both countries were equally welcome to join, which was really the only stance the IGY committee could take. But in the face of the hostility between the two nations, that welcome hardly mattered. IGY could not—would not—proceed with only one of the two countries participating. The whole point of IGY was to foster unity and international cooperation, and IGY was about to find itself in the painful position that if it couldn't proceed with the involvement of either nation, it might have to forgo the involvement of both.

The situation reared its head in September 1955, when Sydney Chapman reported to CSAGI that the communist People's Republic was forming an IGY committee in anticipation of participating in the event. Good news, indeed. Except for the fact that that inclusion was contingent on the *ex*clusion of Taiwan. That didn't seem to be an issue at first, as, after four years, Taiwan had yet to respond to its invitation to join IGY.

Meanwhile, the People's Republic forged ahead with an ambitious program of research that focused on meteorology, geomagnetism, and seismology. The country's scientists reported to IGY that it was proceeding with plans to build twenty-seven observation posts on the mainland, and the communist scientists made a show of attending a regional IGY conference in Moscow in August 1956. A few months later they attended a high-visibility CSAGI conference in Barcelona. There could be little doubt the proud superpower was putting on a show, and it did not go unnoticed.

Apparently shaken by their rival's appearance in Spain, representatives of Taiwan's scientific community contacted CSAGI officials and announced their intention to participate in IGY. The Taiwanese had never received the invitation to submit a program, but they were aware of the event and declared their intention to become an official IGY partner.

On one condition: Taiwan would participate if, and only if, the People's Republic were to be excluded. The People's Republic, meanwhile, would participate if, and only if, Taiwan were to be excluded.

The resulting impasse was the closest IGY had come to an existential crisis, and it was a big one. IGY's mission to remove all political influence and interference from scientific endeavors was teetering on the brink of disaster, and neither of the parties involved was willing to give an inch. "The result might be the wrecking of the IGY," wrote reporter Sullivan.[17]

CSAGI director Marcel Nicolet had to find a ray of daylight among the gathering darkness, but his first attempt was perhaps a little off the mark. He reminded Chu Chia-hua, the director of Taiwan's national scientific academy, that IGY's invitation to join had in fact been delivered to Taipei some four years earlier, in 1952. The insinuation, of course, was that Taiwan had erred in waiting four years to accept the invitation (if in fact it had received one) and had therefore missed some heretofore unknown IGY deadline. Nicolet further reminded the scientist that IGY was a purely scientific event and that the injection of politics into the process was entirely unwelcome.

Tensions rose with the planning for a regional IGY conference to be held in Tokyo in February 1957. The Taiwanese academy requested a formal invitation from IGY to the conference, but the communist Chinese informed the planners in Tokyo that they would be sending their scientific delegation only if Taiwan would be barred from the meeting. Events escalated. CSAGI requested that Taiwan submit a list of its proposed delegation members along with its required program plan. Taiwan responded by asking why the request from IGY was not accompanied by a formal invitation to the conference. CSAGI responded that the members of the CSAGI Bureau were not able to convene to vote on whether to issue an invitation and then insisted that Taiwan assure the IGY officials that they would only be sending scientists, not political operatives, to the conference.

After a series of heated phone calls and cables, Taiwan acceded to CSAGI's request and sent its program plan and manifest of the scientists to be sent to Tokyo. In the relative calm that followed, Chapman felt it would be fair to invite the Taiwanese to the Tokyo conference, because, after all, Taiwan was an ICSU member in good standing and, as such, was free to attend CSAGI conferences.

However, in the same message, Chapman put his thumb on the other side of the scale by conceding that the scientific contributions that the People's Republic of China would make to IGY were "obviously much greater" (and, reading between the lines here, much more valuable) than what could be expected from Taiwan.

Back and forth it went, until the Tokyo conference was suddenly on the horizon. As unappealing as it was to CSAGI and the overall spirit of IGY, it was looking as though a compromise would have to be reached. And then, suddenly, the dam broke.

In the end, tiny Taiwan won the battle of wills, and "thus when the IGY began . . . the nation whose area ranks third in the world was officially not a participant," wrote Sullivan. "Nevertheless, China seems to have carried out all or most of the program originally planned. Satellite-tracking stations were set up across the country, from Sinkiang to the coast. The new observatories went to work and, according to Chinese press reports, the number of people working in the geophysical sciences was expanded from 40—most of them weather men—to 10,000. The roster of weather stations grew more than twentyfold to a total of 1000. Geomagnetic and seismic networks were established, and an oceanographic research vessel fitted out."[18]

The People's Republic of China was the only one of sixty-seven member nations to quit IGY. And according to Sullivan, it was also "the only case in which the IGY was significantly affected by political considerations."[19] But that's not entirely true, as we now know. As the Taiwan/China drama unfolded, there was another political kerfuffle waiting in the wings. Both the United States and the Soviet Union made bold pronouncements in 1956 that they would launch artificial satellites into orbit around the earth as the centerpieces of their respective IGY programs, but at the time only one of those countries seemed to have even the remotest chance of reaching that goal. After all, the United States was the world's technological powerhouse, while the Soviet Union was a backward society of starving peasants and grimy, resentful factory workers building clumsy trucks and tractors that fell apart the moment they were built, right?

Wrong, of course. Very wrong. Only a few months into IGY, the Soviets leapfrogged ahead of the United States and the rest of the world in a show of technological superiority that still inspires profoundly bewildered awe. And the only reason it came as such a shock is that the Soviet rocket scientists participating in IGY made a habit of withholding technical data from their IGY brethren, in direct violation of IGY rules. We'll learn more about that little cold war later; for now, suffice it to say that with everything in place at the dawn of the IGY, science was in for some rude surprises.

A quote from CSAGI director Nicolet comes as close as anything to summing up the mood at the approach of IGY: "Alas, geophysicists are well aware of the fact that, although scientific ideas have greatly influenced human behaviour, they have not yet succeeded in controlling it. In spite of this, the two Polar Years of 1882–83 and 1932–33, and also the IGY in 1957–58 have, together, and in different ways, made important contributions to present-day achievements in geophysics. If one were to try to describe each of these three events by a single word, appropriate choices would be *sublime* for the First Polar Year, *courage* for the Second Polar Year, and *audacity* for the International Geophysical Year."[20]

Interview with Paul Arthur Berkman, PhD

Founder and President, Science Diplomacy Center™; Senior Fellow, United Nations Institute for Training and Research (UNITAR); Faculty Associate, Program on Negotiation at Harvard Law School

Mark O'Connell: Hello Professor Berkman, let's talk about what relevance IGY has to us today. Is that too big a question to ask all at once?

Paul Berkman: It's probably not big enough. Yeah, it's probably not big enough.

M.O.: You've spent a good amount of time working in the Arctic and Antarctica, and were involved in a summit in 2009 to mark the fiftieth anniversary of the Antarctic Treaty. Do I refer to you as a glaciologist?

P.B.: I'm a science diplomat.

M.O.: Can you elaborate on that?

P.B.: My Antarctic background actually is more extensive than my Arctic background. I spent twenty-five years leading expeditions to Antarctica, starting from when I was twenty-two. And spent a year in Antarctica when I was twenty-two, working on an expedition with Scripps Institution of Oceanography. My interest as a science diplomat, is what science means. Okay, what do we do with it? What do we do with our understanding of, of natural sciences, social sciences, and indigenous knowledge? What do we do with it? And that question has preoccupied my life since I was a little boy. And I don't know your orientation to the IGY, but a primary question that I asked when I wintered in Antarctica, scuba diving under the ice every other day, was "Why did the United States and the Soviet Union cooperate continuously in Antarctica, as well as outer space throughout the Cold War, despite the animosities that isolated them everywhere else?" And the answer to that question, which I began forming when I was in my early twenties, decades later became the field of science diplomacy.

I have worked with glaciologists, and one of them whom you would have enjoyed talking with was Charlie Bentley. Charlie Bentley was a young scientist during the International Geophysical Year, and he became famous for determining the thickness of the East Antarctic Ice Sheet that was an order of four kilometers thick. So up until that we didn't even know how much ice was in Antarctica.

M.O.: I've become very familiar with Charles Bentley in the course of writing this book. Why is it important that Bentley discovered the thickness of the ice sheet?

P.B.: You know, the essence of climate change is two phases. There's a cold phase when glaciers grow. And there's a warm phase when glaciers melt. And historically, about 90 percent of the time in the last, let's say, three to four million years, the earth has been in a glacial condition 90 percent of the time. The warm periods, the four periods occur every one hundred thousand years or thereabouts. We're living through a unique one because it's clear that humankind is influencing our global climate. Just in terms of scaling arguments, if you need to know how much water is locked in the Antarctic ice sheet, it would raise global sea level about 120 meters.

So one of the outcomes of asking these types of questions since I was a young lad has led to the development of a theory on informed decision-making, that an informed decision operates across a continuum of urgencies by definition. So one of the challenges, if you think about what a continuum is, it has a start and it has a duration. It's a line of time. And the challenge, in a sense, is the content. So if you think about the IGY absent context, it just becomes a bunch of details. So the context of the IGY is what is fascinating.

Now are you interested, for example, in questions that relate to climate and human influence on climate? Or are you interested in questions that relate to how superpowers interact with each other?

M.O.: Do I have to choose?

P.B.: No, not at all. Because they both fascinate me. So in a sense, if you're thinking about the Antarctic, or the IGY, the reference point is the earth system. And we even began during that period, the journey beyond the earth with satellites. And I'll tell you a story: I was at an Explorers Club dinner in New York a few years ago and sat next to a man named Frederick Ordway, who was an assistant to (former NASA head) Verner von Braun. And I was sitting next to his assistant, J. Frederick Ordway, and he mentioned to me that the United States had the capacity to launch a satellite before Sputnik. Now that flew in the face of everything I had ever heard of the concept of the first satellite being Sputnik. And I asked, how was this the case? And he said that Verner von Braun was under specific instructions from Eisenhower not to launch a rocket into orbit.

So you know, that's, that's a pretty tall change in terms of the history of satellites. And so that, for me, became a rabbit hole. But the journey to discover something is really where the fascination is. That brings me back to the period in Antarctica when I was twenty-two. I came back with this question [of] why the United States and Soviet Union cooperate continuously in Antarctica, originating from the International Geophysical Year? Why did they create the Antarctic Treaty in 1959? Why was the Antarctic Treaty the very first nuclear arms agreement? It was explicitly related to the International Geophysical Year, but the observation was that in Antarctica, the United States and the Soviet Union cooperated because of science, and that was the origin of science diplomacy.

M.O.: How far down this rabbit hole did you go?

P.B.: The concepts of Antarctica, outer space, [and] nuclear weapons began to appear together early in the Eisenhower administration. And the idea was that at the time in the early 1950s, that rockets eventually would carry nuclear weapons. And so the challenge that humanity faced at the time—and still faces, quite honestly—was one of having rockets that can carry nuclear weapons. And so at the time it was satellites that actually brought the Soviet Union into the International Geophysical Year.

Our Joint Chiefs of Staff recognized during discussions in the National Security Council that great prestige, quote unquote, would be afforded to the first in space. But Eisenhower overruled the Joint Chiefs and enabled the dialogue to continue with the Soviet Union during his presidency, including the very beginning discussion about open skies.

The very first open skies discussion coincided with the first U.S. space policy, and the idea of open skies and mutual verification and everything else that went along with it. So it turns out that Frederick Ordway actually gave me the crumbs enough to be able to document that, indeed, [the] United States had the capacity with von Braun to launch a satellite in orbit before the International Geophysical Year. So the question remains, why did the United States choose not to launch a satellite and to be the first nation on Earth to have a satellite? And it goes back to the context of outer space being open and free, that all nations have the freedom of outer space. And that outer space is used for peaceful purposes. And the observation at the time, from Eisenhower, who understood the horror of world war firsthand and understood what would happen if we had a world war with nuclear weapons, which is why his whole concept of Atoms for Peace was explicit in his first inaugural address. And the answer to the question why the United States chose not to launch a rocket.

Once the Sputnik went around the world, once they heard that little beeping sound and as much fear as it put into the allies, and all of those concerned about the Soviet Union, the very first orbit that Sputnik made, Eisenhower was able to claim, "Look, we have a rocket that went around the earth, (because) space is open and free for all to use." And that became the basis for the 1967 outer space treaty that looked at the peaceful use of outer space. So if you go back to the International Geophysical Year and what it accomplished, it enabled the two superpower adversaries to build on common interests, and it enabled them to create the very first nuclear arms agreement in enabled dialogue that has accelerated over the subsequent decades.

MO.: How has the importance of this agreement developed over time?

P.B.: You're working with something that is special to humanity. So the Antarctic Treaty was the first nuclear arms agreement, then as the verification, the inspection provisions in the Antarctic Treaty are among the most sophisticated of any agreement among any nations anywhere in the world, because they allow for any nation, any consultative party to visit any area of the continent without permission. The entire continent is open for inspection. And it was that type of provision that ensured that nuclear weapons weren't being stored in Antarctica. And you can imagine the Antarctic area, there's no indigenous population. There's nobody to complain about testing of weapons there. So the fact that Antarctica became the first nuclear-free zone on Earth is remarkable. And it reflects the need of the United States and the Soviet Union to address matters of survival.

M.O.: What is the role of science diplomacy now?

P.B.: One of the features of being a science diplomat is thinking in terms of common interests, rather than in terms of resolving conflicts. Sure, science diplomacy itself is a language of hope. So my interest isn't to control. It's simply to introduce options without advocacy, which can be used or ignored explicitly, that contribute to informed decisions so that institutions and individuals can operate across a continuum of urgencies.

The IGY is the template as far as I'm concerned, not only because of its global focus in relation to the sun, but because it served as an example of how to bring allies and adversaries alike together. Right. And quite honestly, we're facing exactly the same circumstances with Russia today as we did in the 1950s. And we're creating a new cold war. And so the importance of the IGY, I think, is one of establishing long-term precedent. And I think about the context of the International Polar Year in 1882, that emerged after the Little Ice Age in Europe, that had lasted for three or four hundred years and had created glaciers moving through countries like Switzerland. But the International Polar Year as a process, which began in 1882, is the oldest continuous climate research program humanity has created. And today, if we're emotionally locked in discussions and debates about the politics of climate change, being able to cast the whole process as something that's been ongoing for 150 years, we'll remove the emotion and it'll place the discussion in the right order of magnitude. And if humanity is going to address climate, it's a forever problem. It's not something that we can resolve because of an accord in 2015. It's a process that humanity requires forever. And that process actually began in 1882. And by creating that history, that legacy, it enables humanity to think over timescales which are anathema to the way the world works.

And in a sense that capacity of science to operate on a planetary scale has always been a necessity. Always. And we are evolving as a civilization. And the sense is that, you know, just in the most basic thing, it's not something that requires education, even just simple observation. That when any of us make decisions about anything in our lives, we have methods to make those decisions, to give us the ability to understand change, to understand how to address change; those methods come from the natural sciences, the social sciences, and indigenous knowledge, which have been evolving in our world over millennia. And we as a world are in a different place now than we've ever been, particularly because of advanced technologies and industrial capacities. And our challenge is no different than it was in the 1950s.

It's one of survival. And our responsibility, all eight billion of us, is to avoid another period of global conflict. And if you can help it along those lines, you've done something important.

We are just in our infancy, and like infants, we bump into things and we make mistakes. And the challenge is that we have to survive beyond our infancy in the absence of an adult to show us the way. And the adult in the discussion is science. The International Geophysical Year shows that the ability to operate beyond the boundaries of nations, with individuals and institutions, that capacity exists with science.

CHAPTER 3

First the Sun

What could be more audacious than to begin with a study of the sun? It is somehow fitting to award it pride of place here, because, as stars go, it is not at all remarkable in the overall scheme of things. Some 4.5 billion years ago, all across the universe, uncounted billions of vast balls of gas and dust suspended in space began to collapse of their own mass and weight and eventually agglomerated into a number of solid forms. For our purposes, we will concern ourselves only with the forms of gas and dust destined to become our birthplace and our home. Those forms slowly took on what would become one of nature's favorite shapes: a sphere. Most of the spheres emerging from the collapsing cloud became our home world and our neighbors, the planets, and at least some of their respective satellites, but one in particular became much more. The central sphere captured, by pure luck, just the right types of that gas and dust in just the right proportions, subjected to just the right amount of pressure, to combust, to create heat and light, and to keep itself burning from the inside out for billions of years. But what good is light if there's nothing on which it can shine? The universe had an answer to that question and made sure that those planets and their satellites would remain where the power of the sun could always hold sway over them.

Now classified as a yellow dwarf star, our sun has reached middle age, but it still has a lot to offer Earth and the other planets for billions more years as it slowly swells up to become a red giant and then ulti-mately compress into a white dwarf. Not enough mass to create a black hole, alas. While it may be unremarkable in astronomical terms, it is to us on Earth (and elsewhere in the solar system) a most remarkable and necessary bundle of explosive energy.

Here on Earth, we learn very quickly that nothing can burn with-out the flames being fed by oxygen. But on, and in, the sun there is no oxygen; the fires of the sun are instead fueled by hydrogen, the lightest and most plentiful element known to us (oxygen enters the picture much later). The hydrogen atoms in the sun are continually, endlessly smashing into each other to create helium atoms, the act of which in turn generates the heat that powers and propels the entire solar system. The sun started out with a 10-billion-year supply of hydrogen to burn and at this moment has about 6.5 billion years' worth left in the tank. It staggers the imagination.

*   *   *

On his inaugural IGY journey, J. Tuzo Wilson also started with the sun. As the vast machinery of IGY rumbled to life around the globe, he traveled from his quiet Ontario cabin to the ghost town of Climax, Colorado, elevation 11,342 feet, to get acquainted with his first solar observatory. This abandoned mining town was the site of the High Altitude Observatory (HAO), operated by the National Bureau of Standards (NBS) and the University of Colorado, and one of 120 such observatories keeping their eyes unerringly trained on the sun for IGY.

If you, like me, have ever wondered what the NBS does, much less why it was created at all (especially since we already had the Of-fice of Weights and Measures), you'll enjoy the rest of this paragraph. The NBS, commissioned by the U.S. Congress in March 1901, has been described as the very first geophysical research organization in the country, as it was involved in such missions as developing high-altitude balloons for weather research, understanding what happens when we "ground" an electric circuit, and even inventing the atomic clock. The

NBS was established to meet the growing needs of industry, commerce, and technology to share rock-solid manufacturing and building standards, so that a nut cast in Kansas or California, for example, would screw on perfectly to a bolt cast in South Carolina or Illinois. The need for standards of this kind was hammered home by a fire in Baltimore in 1903. Fire brigades made it to the scene of the fire, many from neighboring cities, but the dismayed firefighters discovered that the out-of-town hoses wouldn't fit the Baltimore fire hydrants. The firemen could do nothing but watch as the fire consumed more than fifteen hundred buildings and homes. In the aftermath, the NBS agents sifted through fire-hose couplings of six hundred different designs to select a national standard and avoid another Baltimore fire debacle.

It made sense, then, for the NBS to involve itself in the activities of IGY, as so much of IGY's research depended on the use of identical instruments behaving identically at the same identical time all over the world. And here was Wilson getting his first real look at how that would work. The first of the observing "eyes" he encountered on a guided tour of the NBS facility was the wire dish radio antenna, tuning in radio waves from the sun in its massive metal webbing. The second was an optical telescope with a red filter that enabled the astronomers to view solar flares by the red light they gave off, which automatically took a photograph of the resulting light show every minute. The third device was a clever telescope called the Lyot after the French astronomer who invented it. Lyot's instrument was an ordinary telescope with a metal disc placed in the tube in just the right position to blot out the disc of the sun, creating a custom-made solar eclipse for the viewer and enabling observations of solar prominences and flares around the sun's edge. So simple, so clever.

While getting a look at the sun through the Lyot, taking in his first view of wide swaths of solar flame leaping from the sun's surface, some of them millions of miles across, Wilson likely found himself humbled and unnerved by the staggering size and violent nature of the sun.

*   *   *

The solar scientists of IGY had a sizeable list of investigations to conduct, simply because the sun does not keep its explosions to itself. It is constantly shedding CMEs that reach out across space to affect the earth, bathing us in light, heat, and radio waves that can create things of great beauty, such as the aurora borealis, as well as wreak havoc with transportation, communications, and power grids. It was known at the time of IGY that a portion of the sun's energy—namely, light, heat, and weak ultraviolet rays—was constantly bathing the earth, keeping us alive and sometimes giving us great tans. At the same time, it was suspected but not known that the sun was sending much more dangerous and interesting energy our way, but that energy was being blocked by our atmosphere, making it impossible to detect.

The solution to the problem? Get somebody or something—maybe a balloon, maybe a rocket, maybe a hybrid of the two, cleverly named a "rockoon"—aloft, and get it or them to ascend clear of the earth's atmosphere—where it would be bathed in those dangerous rays for a time, getting what Wilson described as "a clear, unfiltered view" of the sun—and secure as much data as possible from the observations made before inevitably losing altitude and returning to Earth, either gently or, more likely, not so gently. At the time we were able to get rockets up into space, or close to it, but we had problems getting them to stay there. In 1957 a rocket was a rocket: it went up, it went down. A satellite was a different animal altogether: a satellite went up and stayed up, usually for a pretty long time. We'll get to that later.

*   *   *

"It now appears that space, at any rate around the earth and sun, has a 'climate' as changeable as that on earth's surface," Wilson noted in his book *IGY: The Year of the New Moons*, adding that Earth "is continually shaken and buffeted by a bombardment from the sun of electro-magnetic waves of many frequencies, called radiation, and on high-speed atoms and fragments of atoms which strike the Earth's atmosphere in tenuous blasts of 'solar wind.'"[1] Solar scientists at the time of IGY considered it vitally important to understand (a) with what kind of energy the sun was bombarding us, (b) from what part of the sun that energy was coming, and (c) how that energy affected the earth and its inhabitants.

There are, in fact, three distinct types of observable phenomena arising from the solar inferno. Determining how these three phenomena interact with Earth and with each other has for years been a field of intense interest among astronomers.

- Sunspots are brilliantly incandescent and appear to us as black spots only because they are cooler and darker than the rest of the sun's surface. They are marked by intense magnetic activity, often appearing in pairs straddling the sun's equator, but only between the solar latitudes of 35 degrees north and 35 degrees south. One sunspot from each pair always has a negative polarity, and its counterpart always has a positive polarity, and these polarities flip every eleven years. Harold Fraser remarked in 1957 that more than this "we can only guess" and that "nobody knows" how sunspots really work. (That has since changed, but not to the extent you might expect.)
- Solar prominences are plasma loops that connect two paired sunspots. They show themselves around the rim of the sun during solar eclipses, erupting through the sun's corona, or outer shell, and then falling back to the sun's surface. They follow the same eleven-year cycle as sunspots, but not the same spatial patterns, often appearing far north or south of the sun's equator.
- Solar flares (now known as CMEs) are far larger and far more combustive than prominences. They erupt quickly, in a matter of a few minutes, spreading out to cover millions of square miles, and then fade away after an hour or so. Flares always form over sunspots, but not over every sunspot. They also bombard the earth with stellar forces that even now challenge our understanding.

The sun put on a couple of its own special fireworks displays to mark the start of IGY and put smiles on the faces of a lot of scientists around the world. First, on June 28, 1957, only a few days from IGY's kickoff date, a Soviet tracking station near Moscow called in a report of a large solar flare to the World Warning Center, the IGY's clearing house for daily solar weather reports in Fort Belvoir, Virginia. Thirty-six hours later Europe was hit with a massive radio blackout. Then, on July 1, the first day of IGY, a solar flare was reported by the Sacramento Peak Observatory in New Mexico. Just as in Europe a few days earlier, a radio blackout hit when the eruption reached the earth's atmosphere.

"Scientists in every region of the earth took simultaneous measurements of the solar effects. Not only did radios go dead, but extraordinarily bright auroras were seen," Alexander Marshack reported. "U.S. IGY scientists fired a rocket 75 miles into the air off California to catch the incoming radiation."[2] It was a picture-perfect, camera-ready debut of some of the most sophisticated resources and processes that IGY had to show off to the still-skeptical world, and it made headlines around the globe. For the first time, millions of people around the world could understand the "why" of IGY.

The show continued throughout the duration of IGY, as *Life* magazine reported: "As it turned out, the sun staged its most explosive show in 200 years. Twenty-one times a World Warning Center in Virginia radioed 800 world observatories to expect especially turbulent bursts."[3]

Getting a handle on what the sun's atmosphere consists of and what it does to the earth can be difficult, in part because so much of what the sun sends our way is literally invisible. To understand the mechanisms involved, we must start at the lowest layer of the sun, just as the IGY scientists did. By viewing the disc of the sun with the Lyot telescope as Wilson did on his Colorado trip, scientists were able to detect three distinct layers.

- The photosphere is the lowest layer, and it is visible—most of the sun's energy emerges from here.
- The chromosphere covers the photosphere. It is not so visible, is not so understandable, and can be observed only during a solar eclipse or when looking through a Lyot.

- The corona, the outermost layer, rests above the first two, so we might think it would be the most visible to us. Yet it, too, is seen only during an eclipse, or when looking through a Lyot, and it is an uncomfortable three hundred times hotter than the chromosphere below it.

The sun is a violent body, but also consistent, at least when viewed from the surface of the earth. But then something curious happens. "About every 11 years the sun suddenly goes wild," *Life* reported, "shooting out incandescent prominences that rise either like flaming hedges or like arches, jets and loops."[4]

Galileo himself was among the first to perceive and study the dark spots that appear on the face of the sun, in the 1600s, using the recently invented telescope to bridge the visual distance that existed between planet Earth and our star. At the time it was held that the spots were separate from the sun, but after many observations Galileo declared that they were actually a feature of the sun's surface and that even the smallest sunspot was many times larger than the earth. This did not win Galileo any popularity contests, as the sun was widely considered a pristine celestial body, free of blemishes and imperfections. Despite opposition from the Catholic Church, Galileo continued his studies, and over time he learned a great deal about the sun and its odd spots. He found that the movement of sunspots across the face of the sun indicated that the sun rotates just as the earth does. He learned that it takes the sun twenty-seven days to make a complete rotation and that there seemed to be no pattern to the appearance of sunspots: one day there might be dozens, the next day none at all. Sometimes several years would pass without any sunspots appearing.

Galileo would have been astonished to learn that by the twentieth century we could keep the entire disc of the sun under tight scrutiny twenty-four hours a day and thus never miss a single sunspot. "During the 18 months of the IGY, the sun is under observation or patrol every minute from some place in the world," explained Alexander Marshack in his 1958 book about IGY, *The World in Space*. "When the sun disappears below the horizon for one astronomical observatory, it is already visible to another."[5]

*   *   *

The *Life* magazine article referred to earlier was the last of four articles in a series titled "New Portrait of Our Planet," which ran in late 1960. Its aim was to educate the populace about not just the "why" of IGY but also the "what," the "how," the "where," and, in some cases, the "when." After publishing three articles illustrating IGY's impact on Antarctica, the earth's atmosphere and magnetic field, and its water and winds, the magazine turned its attention to "The Sun's Awesome Impact."[6]

The major thrust of the article was that IGY had proven beyond all doubt that the sun had an atmosphere much like the earth's, and that the sun's atmosphere actually enveloped Earth's. It was a strange concept to think that our planet was wrapped in one atmosphere that was in turn wrapped in another, much larger, much more volatile atmosphere.

Through a series of full-color paintings both intricate and vast, *Life* showed on its oversized pages how the earth and the sun interacted in space, how the scientists of IGY discovered the regions in the sun from which each type of radiation originated, and what those rays do when they slam into the earth.

What we experience as light is a very narrow band in the center of the spectrum. This light originates from the sun's photosphere and reaches the earth eight minutes later. Beyond the sunlight, just outside our range of vision, are radio waves with longer wavelengths that originate in the corona and are detected by radio telescopes like the one at Climax. Just beyond these rays are infrared rays emanating from the sun's photosphere, keeping us warm and often triggering the creation of storm clouds.

On the other side of the spectrum are ultraviolet waves with shorter wavelengths than those of visible light. These originate in the sun's photosphere, and they are deadly. In small amounts, they can give us a painful sunburn, but at larger exposures they will kill us. The fact that ultraviolet rays are not killing us comes down to the existence of the ozone layer sixteen miles overhead and its ability to transform ultraviolet waves into a more benign element we call oxygen.

Beyond the ultraviolet are rays that ionize our upper atmosphere, colliding with molecules and atoms and giving them positive electrical charges called ions. While they create ions out of nitric oxide and nitrogen, ultraviolet rays are playing a huge role in global communications, as discussed previously. It's the ionized atmosphere that causes the mirroring effect enabling us to "bend" straight radio signals around the sphere of the earth.

But what nature giveth, nature taketh away, as further along the spectrum we encounter a type of wave that inhibits radio communication. The bad actors in this situation are the X-rays emanating from the sun's outermost layer, the corona. Some X-rays ionize oxygen and nitrogen molecules, and some make appearances in solar flares; if they exert enough downward pressure on the earth's atmosphere, they can cause those troublesome radio blackouts we've already encountered.

A December 1965 report gives us a comprehensive look at the solar activities performed for IGY, and it's impressive to see what work was to be done, how it was to be done, by whom, and for what purpose. The report begins with a summation stating that "a large part of the U.S. program involved continuation of current work with more emphasis on patrol and on prompt reporting of summary results to workers in the disciplines participating in the overall IGY work," which suggests that portions of IGY's work were already being done pre-IGY, but perhaps with less focus or discipline. Perhaps the new world of IGY caused scientists in the United States and beyond to sharpen their work habits and expand their reporting? In any case, the panel on solar activity specified a need for (1) more extensive observations at existing solar observatories, (2) flare patrols to include both photographic and visual observations, (3) more extensive daily measurements of sunspots, (4) measurements of the quantity and quality of solar radiation, and (5) radio observations of the corona.

Further into the report, we find that the U.S. solar scientists were adjusting their plans right up to the last minute, adding "major extensions" to the program, and that the HAO at Climax was the central hub of the research operation. These extensions were as follows:

- installation of a solar patrol station in Hawaii, with funds to the University of Hawaii to construct special photographic flare patrol equipment;
- equipping more stations to observe the radio frequency spectrum bursts and outbursts, located at the University of Michigan at Ann Arbor and the Radio Astronomy Station, Fort Davis, Texas;
- equipping three more stations to observe sudden enhancements of atmospherics and sudden cosmic noise absorption as a flare-warning device, to be installed at the Observatory at Boulder, Colorado; at Sacramento Peak, New Mexico; and at McMath-Hulbert Observatory, Pontiac, Michigan; and
- additional patrol work on mapping solar magnetic fields performed at Mt. Wilson Observatory in Mount Wilson, California, using specially developed instruments and electronic-optical techniques.[*]

"The International Geophysical Year, might logically be called 'The Year of the Sun,' for the IGY was called in 1957–58 so that the effects of the sun on the earth could be studied during a peak in sunspot activity," wrote Marshack in *The World in Space*. "The attempt to find the secret of these spots has gone on for a century."[7]

---

[*] You may have noticed something of interest in these brief report notes on the IGY solar program: 1957–1958 was a good time to be in the scientific instruments business. In addition to the items on this short list, IGY installations around the world, flush with funding, were ordering new spectroscopes, cameras, and telescopes of all kinds with which to do their solar observing and analysis. And it wasn't just the solar scientists; researchers in every branch of geophysical science around the world were taking advantage of an unprecedented influx of funds to order new instruments and materials in droves. It must have been a heady time for the instrument craftspeople and salespeople as they watched their sales charts climb and climb like an unleashed rockoon. And while there is no record of the overall amount budgeted for and spent on these devices, a look though the final report of the U.S. IGY efforts suggests that companies supplying IGY scientists with the tools of their trade must have had a very hard time keeping up with the dramatic increase in demand.

Sunspots are hardly the only mystery associated with our sun. I admit that even after spending many long days at my desk trying to understand how our sun functions, I am still rather baffled by how it all works. So, to shed some more daylight on the topic, I did what any confused writer would do: I sought out some experts who could make sense of the myriad and multitudinous chemical reactions going on up there. My first call was to Dr. Emily Mason, a research scientist at a private lab in San Diego, where she conducts remote observations of the sun, primarily in the "extreme ultraviolet regime." Dr. Mason's work involves "complex 3D plasma simulations," which, honestly, sounds like something that I might have cooked up for one of my *Star Trek* episodes. Which made me feel a little more at home.

When I asked Dr. Mason to elaborate and elucidate, she dove right in: "My research focus ranges across quite a few of the phenomena observed in the low solar atmosphere (corona), from the cool, large-scale dark regions known (somewhat melodramatically) as coronal holes, to the super-heated knots of magnetic fields (ironically mundanely) called active regions. Everything that affects the earth and the other planets in our solar system—as well as having potentially catastrophic effects on astronauts and robotic missions—starts in the lowest reaches of the corona, but it's such an inhospitable and elusive region to try to learn about."

Tell me about it.

"It's a truly understated challenge in astrophysics," she went on. "The sun is so tantalizingly close, it's one of the only astrophysical bodies we can resolve, you don't even need a telescope, yet it still maintains so many mysteries."

Although it is outside the sphere of her expertise, Dr. Mason was familiar with the rough outlines of IGY, but in her case it was through the lens of the Cold War. In her memories, IGY was about polar missions, radiation belts surrounding the earth, and early rocket launches from "extreme latitudes," which she told me are still a part of her fieldwork on her solar projects. To Mason, IGY research was primarily remembered as a component in "near-Earth space exploration and development," which must undoubtedly embrace the sun.

I was curious to get Dr. Mason's take on the sun's atmosphere and how it envelops the earth's atmosphere and acts on it, but she politely informed me that there were "a few things to clear up" about my inquiry.

"In the sense that the same particles that leave the sun bathe the earth and the rest of the solar system, it's still the sun's atmosphere," she explained. "However, speaking from a scientific and space weather perspective, it's only the atmosphere until the solar wind becomes supersonic—at only around 1 percent of the distance to the earth—at which point it becomes the solar wind. That distinction between the corona and the solar wind was established in a famous paper (by solar standards) in 1958.

"But we also knew that the sun had its own atmosphere since at least the late 1800s, based on early spectroscopic eclipse observations. Furthermore, it was a debated but increasingly-agreed-upon fact that eruptions from the sun could have an effect on the earth since the Carrington Event in 1859 [more on that spectacular solar event later in this book]."

Having set the historical record straight, Dr. Mason assured me that she was through with what she called her "curmudgeoning," and she was happy to continue serving as my solar tour guide. First, she referred back to her earlier comment about near-Earth space exploration, explaining that the early space-age discoveries that came out of IGY were "huge" and created a unique opportunity for solar scientists to learn how to predict space weather, an idea she described as "tempting and still currently beyond reach."

She cited the aurora borealis (aurora australis in the Southern Hemisphere) as the classic example of the interaction between the earth's atmosphere and the solar wind. The solar wind creates pressure that deforms the earth's magnetosphere (the magnetic envelope that keeps our atmosphere in while keeping cosmic radiation out). This disturbance in our magnetic field energizes the gases in our atmosphere. These emit light to "cool off," and that light becomes visible to us as the aurora. "If you look at the aurora, you're seeing the pattern of the solar wind imprinted on our magnetic field," Mason explained.

She then cited radio blackouts, which occur when solar flares erupt. If a flare is strong enough, it can emit ionizing radiation that can penetrate the earth's ionosphere. The lower level of the ionosphere makes high-frequency radio communications possible by bouncing the radio signals around the curvature of the earth.

Solar events can affect the earth in two ways: first by emitting the aforementioned radiation, and second by emitting CMEs. "CMEs are huge clouds of magnetic field and plasma that are thrown out of the sun's atmosphere but retain their connection to the sun," Dr. Mason explained. "If the field is oriented opposite to our magnetosphere, Earth experiences a geomagnetic storm, which can affect GPS, power grids, cellular communications, and other systems. It is next to impossible, given current technology, to predict which way the CME's field is oriented before it hits."

Depending on their strength, solar flares can cause CMEs or radiation storms, Dr. Mason said. "The radiation effects from flares are more damaging for objects and people in orbit; we've actually been incredibly lucky that none of our astronauts have ever been on the moon or in orbit during a severe radiation event. Any one of the Apollo missions could have ended in tragedy if they had happened to be caught out of the earth's magnetosphere during a major flare, as the moon dips in and out of our magnetic sheath. Satellites are very sensitive to electrostatic charges, and the radiation from a solar flare can literally shock them into space junk. The additional ionization of the top layers of our atmosphere caused by the flare can also increase drag on spacecraft, sometimes catastrophically (as Starlink found out [in 2022])."

At this point I had absorbed enough to surmise that ionization was a pretty important part of the sun-earth relationship, so I had to ask Dr. Mason a real beginner's question: Can you describe the process of ionization?

"Sure," she said with no hesitation. "Atoms and molecules have electrons that buzz about the outer boundaries in complex patterns. These electrons can convert between different forms of energy in fascinating and efficient ways. Generally, if an electron absorbs energy, it moves farther away from the nucleus of the atom to which it 'belongs' (this belonging concept is tenuous at best in molecules, before you even get into quantum theory). However, if bombarded with enough energy of the right type, it can move so far out that it becomes free of the atom's influence entirely. It's kind of like a rocket having enough energy to reach escape velocity. At that point, you're left with a free electron and a now-net-positive atom/molecule, which is commonly referred to as an ion."

That cleared things up quite a bit for me, aided immeasurably by Dr. Mason's image of a rocket reaching escape velocity. But I felt there were still gaps in my solar education—like, what happens to that electron after it's ionized?—so I looked up one of our preeminent solar scientists and educators, Dr. Ryan French, Brinson Prize Fellow at the National Solar Observatory (NSO) in Colorado Springs and author of *The Sun: Beginner's Guide to Our Local Star* (2023).

"Specifically, I investigate solar flares," Dr. French told me by way of introduction. "So, I'm an observational guide, where I use data from telescopes in space and on the ground to try and work out why solar flares happen, and how maybe we can predict them eventually."

I launched into our conversation with a decidedly earthbound issue: How do I make sense of the space weather app I just downloaded on my phone? "Every couple of days I'll get an alert like this one I got yesterday," I told him. "It's a Solar Flare alert informing me that earth-orbiting satellites have just detected an MIT M 1.5. Class solar flare at 11, January 2024, at 17 hours, 52 minutes UTC check space weather .com for updates.

"These alerts I get I find really exciting because they're like, oh, there are interesting things going on in the sky that very few of us are aware of. But then I think okay, but what does that mean? So, earth-orbiting satellites have detected a solar flare? What does that mean to me? Just how does that affect me? Or is it just sort of a curiosity like, 'Oh, that's cool to know.'"

Dr. French explained very patiently that these alerts were being sent to my device by a satellite run by the National Oceanographic and Atmospheric Administration (NOAA), which has an X-ray sensor that points in the direction of the sun.

"We essentially define the magnitude of the size of a solar flare based on how many X-rays it kicks off," he explained. "And the sun undergoes a cycle, I don't know if you've come across this in your research, there's a solar cycle, which is eleven years long. And so this eleven-year cycle increases and decreases the amount of solar activity. So, at the peak of the solar cycle, which we call solar maximum, there are lots of solar flares happening all the time. And at the minimum of this solar cycle, there's nothing happening on the sun at all, basically, no sunspots, no flares, nothing. But we're actually approaching solar maximum this year.

"And this year, they're popping off, we've had like three or four this week, and I'm sure you've seen the alerts come through. So, flares do affect us. In order for it to be a flare big enough for a regular person to actually notice [it] would have to be really, really big. But even small flares, and flares, x one class flares, for instance, they cause these things called radio blackouts."

Dr. French explained that the strongest coronal mass ejections, the strongest eruptions, come from the strongest flares. These flares are actually packets of intense energy combining plasma and magnetic fields that are hurtling away from the sun. If that flare were to reach the earth, it would cause a geomagnetic storm.

"So, if you have a satellite in space, or say, a power line, or something like that, and you have this thing coming from the sun with a magnetic field, and it passes over the conductive satellite or the power grid, you essentially create a current rolling this thing traveling past and you create current in the system," Dr. French explained. "And if that current is big enough, you can cause overcharging on the satellites.

You can overload our grids in the worst cases, and disrupt other things like railways and things like that. That's the worst-case scenario."*

According to Dr. French, this worst-case scenario could involve the loss of one in five of our satellites, seriously disrupting telecommunications, weather monitoring, radio communications, the GPS network, and more, creating all manner of havoc on the earth below. Power grids are also vulnerable, especially grids that are old, as so many are, and centralized grids that must send their power over great distances. Both regional and localized blackouts could be the result of an overloaded grid.

Don't blame the solar flares, though. The larger eruptions that accompany the flares and find themselves on a one-way voyage to Earth are to blame.

So it turns out the space weather alerts on my device would not represent a direct threat to human life. "It wouldn't be apocalyptic," Dr. French assured me. "It wouldn't send us back to the Stone Age, as some people like to say, but it would be a large economic disruption. That would probably ground flights for a couple of weeks. And you'd have to replace a bunch of power grids, or we launched a bunch of satellites, so it'd be a large economic disaster.

"Maybe I over-answered your question there," Dr. French admitted. That is precisely why it is so delightful, even entertaining, to me to listen to a scientist explaining his or her work.

"But okay," I said to Dr. French, "so this may be a dumb question. But when the sun ejects plasma, when these phenomena occur that strike the earth, is this shooting out in all directions from the sun, and the earth is just one tiny part of what is getting exposed to these rays or these particles, or is it somehow literally pointed only at the earth?"

Again, Dr. French answered with great patience, first informing me that there are no dumb questions, and then explaining that these waves of material from the sun actually go out in all directions, so the earth is in the path of only the tiniest portion of the emissions.

---

* I love writing about worst-case scenarios, to be honest.

In hopes of receiving some further "over-answering," I held back as Dr. French continued: "Sure. So, sunspots are these dark patches in the surface of the sun, which are called the photosphere. That's the layer of the sun that we see, and some regions of that layer have very dense concentrated magnetic fields. And this magnetic field, it's so strong, that it basically repels any plasma from around this magnetic area from flooding. So, because of that plasma inside the strong magnetic field, it basically becomes isolated from the rest of the surface. So, because it's sort of isolated, it cools down because it's not connected. But the hot under layer of the sun sort of cools down to lower temperatures. And that's why we see it as dark because it's colder, basically; it's about 2,000 Fahrenheit colder than the rest of the sun's surface. So, these are strong areas of magnetic field, and then above these regions, you have these active regions in the atmosphere where you find that strong magnetic field. And that's where you get the flares and eruptions."

The more Dr. French's descriptions of our knowledge of the sun today related to what was known at the time of IGY, the clearer was my view of how this all tied into the solar research conducted during IGY. I asked whether he agreed that one of the major accomplishments of IGY pertaining to solar physics was the determination that the sun actually had its own atmosphere, and that that atmosphere also enclosed and influenced the earth's atmosphere. Having already posed that question to Dr. Mason, I was pretty sure of what his reply would be, but his answer surprised me.

"I don't know much about the politics around that time, but certainly it was in the 1950s that we discovered the corona of the sun, the atmosphere of the sun," he said. "This partly came from a man called Eugene Parker, who proposed something called the solar wind. He suggested that there should be a constant stream of plasma flowing out from the sun, just continuous plasma coming out. And he was ridiculed for this theory."

According to French, no respectable solar physicist at the time believed a solar wind was possible, but Parker persevered. After all, it had to be significant that when comets circle around the sun, no matter where they are in their orbits, their tails always stream away from the sun; clearly, *some* energy or substance was being continuously propelled out from the sun in all directions.

"You must understand how unbelievable this sounded when he proposed it [in 1957]," said a University of Chicago professor of astronomy and astrophysics who worked with Parker at the time. "That this wind not only exists, but is traveling at supersonic speed! It is extraordinarily difficult to accelerate anything to supersonic speeds in the laboratory, and there is no means of propulsion."[8]

Only a few years later, when NASA spacecraft *Mariner II* took readings on its journey to Venus in 1962, the results were unambiguous. "There was the solar wind, blowing 24/7," Parker said.[9]

I am pleased to report that NASA honored Dr. Parker by naming a pioneering solar spacecraft after him in 2018. The Parker Solar Probe was launched from Cape Canaveral, with Dr. Parker in attendance, on August 12 of that year, beginning a years-long journey to the sun's two million degree corona. In 2021, the Parker Solar Probe entered the sun's corona, and it continues to venture deeper into the solar atmosphere with every orbit, adding immeasurably to our knowledge and understanding of the sun.

The outcome of the solar wind debate swung open the door for a new era of solar research based on the newly acknowledged fact that the sun had an atmosphere that affected the earth. "Certainly, we're still within the sun's magnetic field and its wider atmosphere," Dr. French said. "There's a constant stream of particles from the sun flowing past us. And I would say the earth's magnetic field is within the sun's magnetic field, and so the earth's atmosphere is then well within the earth's magnetic field."

"Mars used to have a magnetic field about a billion years ago," he continued. "Mars used to have a very thick atmosphere, used to have running water, things like that; we see evidence of rivers and lakes and fans and stuff on Mars. And (at some time) it lost its magnetic field. And then very slowly, the solar wind stripped away Mars's atmosphere until now it basically doesn't have an atmosphere anymore.

"Earth gets its magnetic field from its solid inner core, because the earth's rotation causes it to spin around that inner core. Mars's liquid layer at its core, the whole thing crystallized. So, it became solid, and it no longer had that dynamo energy source. Our Earth's liquid core is very, very thick, with no sign of crystallization. So, what happened to Mars, it's not going to happen to us."

When I pressed him on the matter, he explained that the earth might lose its magnetic field in ten billion years if the sun didn't die before then. "I wouldn't worry about it," he told me.

"Do you remember back in I think it was 2010 when there was an eruption of that volcano in Iceland?" I vaguely remembered it, and Dr. French gave me a refresher: "It was a volcanic eruption in Iceland. And the blooms from the volcano blew across Europe. And flights were grounded for about two to three weeks in the whole of Europe. And some people got stuck on holiday and couldn't come home. And then following that, the UK, Europe, US did a big review on the other natural disasters that maybe had been omitted in their planning, natural disasters that people hadn't considered, hadn't thought of one prepping for. And in the UK, at least, where I'm from, that's when they first decided in 2011, there's this thing called Space Weather. These solar flares and eruptions from the sun could have a similar effect on aviation. And we've not planned for this at all. So that's when the UK started doing space weather forecasting and it started pumping more resources into science. And I think a similar thing happened over here in the US. So the whole grounding of flights was the initial reason that they started worrying about this stuff from the sun to begin with. And there are wider effects in that too, but it started with the aviation industry."

My talks with Drs. Mason and French went a long way toward raising my comfort level when writing about the sun, which seems so simple but is so intensely complicated. Learning that even in the twenty-first century these two brilliant researchers are still puzzled and intrigued by what the sun is and what it does gave me a new appreciation for how profoundly difficult it must have been for the solar physicists in the 1950s to understand what they were dealing with, measuring what must have seemed at times like the unmeasurable.

"I think the most common thing that I try and point out is that the sun does stuff," Dr. French added, taking my solar education in a fascinating new direction. "It's easy to see the sun in the sky, right, and be like, 'Oh, it's just always there, doesn't do anything.' I think as a child, I never considered that the sun did stuff. But the sun's always changing all the time. And, you know, it produces stuff. But not only do we need to care about the sun for the reasons of safety and convenience—because these things can disrupt us, our technology, things like that—but we can also learn so much about the sun, like . . . the sun is a star at the end of the day, right? So, by just watching the sun and learning things about the sun, we learn about stars and our wider place in a universe made of plasma."

"The sun is a star at the end of the day." I confess I was surprised by Dr. French's profundity, but his words made me realize something important: we are already among the stars. We don't have to travel to see a star, interact with a star, study a star. Thanks to our sun, we are among them. When I thanked Dr. French for sharing his time and wisdom, he had one more over-answer to share: "The sun does stuff, and it's vital that we understand what it does and why. Over eons of history the sun can change size, slightly change temperature, too, and there have been cases in history where the sun has significantly changed climates. The thing is the timescales for those processes are thousands of years, tens of thousands of years for significant evolution of the sun.

"The current global warming that we're seeing is absolutely due to the change of carbon dioxide in the atmosphere. And the levels of carbon dioxide, it basically just change[s] the amount of sunlight that is reflected and absorbed, just changes that ratio. So yes, the sun can influence climate, but the sun is not the driver of the climate change we're seeing right now."

So, what is?

*   *   *

More about the sun: Lest the reader think that IGY was concerned only with high-profile, headline-grabbing research and discoveries, I'd like to end this chapter with a brief summary of the objectives and results from some of the lesser-known, smaller surprises of the IGY solar program, taken from the 1965 *Report on the U.S. Program for the International Geophysical Year*.

Summary of Solar Activity Program

The U.S. Program in solar activity included the basic patrols carried on by existing solar observatories supplemented by an intensive photographic patrol for solar flares with new instrumentation and specialized studies of flares and other unusual solar phenomena. Thus, the program contributed to the 24-hour watch on the sun during the IGY and provided detailed astrophysical data on many of the solar events which may be associated with the terrestrial phenomena studied in the purely geophysical disciplines. Projects provided for the procurement, installation, and operation of solar activity equipment and for the collection, reduction, analysis, and publication of data.

Line Profiles of Flares

Objectives: This project at the High Altitude Observatory provided for the modification of the Climax coronagraph and associated spectrograph to permit the taking of rapid-sequence spectra of solar flares. Thus a "map" of the flare is obtained in terms of intensity of H-alpha and other emission lines which would enable the study of the morphology of flare initiation and development.

Results: A great number of extremely interesting spectra of solar flares were obtained by the Climax observers with these instruments; some, while still developing the observing procedures before the actual start of the IGY. Analyses were conducted toward determining the physical conditions of flares.

White Light Corona—Instrumentation

Objectives: This project at the High Altitude Observatory provided for the design and construction of an instrument (white-light corona photometer, or K-coronameter) to be used to measure and record the brightness of the polarized component of light scattered by free electrons in the electron, or K, corona, as distinct from the emission corona. Such observations are important to the study of the physics of the sun and may elucidate those solar phenomena which result in various terrestrial events. Routine observations had never been attempted previously.

Results: From some fifty-five days of usable observations during IGY, it proved possible to develop models of the electron corona above the quiet disk, the polar regions, and active regions. The last of these turns out to depart from the spherically symmetrical model and was compared with models of the radio corona. A comparison of the 21-cm picture of the sun and the intensity of the K-corona showed striking resemblances between regions of enhanced 21-cm radiation and regions of enhanced electron density, and suggests that the enhanced radiation is of thermal origin from regions of high density. It is expected that continued observation of this kind, along with more detailed radio and coronal spectrographic observations, will improve understanding of the physics of active regions.

Improved Coronal Line Emission Studies

Objectives: In response to the needs of the various programs of the IGY the High Altitude Observatory made numerous improvements in its coronal emission line studies. Intensities were reported on an absolute intensity scale, and were measured from the films photoelectrically, by a sound procedure of photometric analysis. For some regions, height gradient studies were made, and also coronal line profile analysis. Attempts were also made to develop more meaningful indices of coronal activity, in an effort to have them more accurately represent the integrated emission of the higher temperature lines of the corona.

Results: A set of $\lambda 5303$, $\lambda 6374$, $\lambda 5694$, and $\lambda 5445$ coronal line intensities, measured on the same absolute scale and relatively independent of observing conditions, have been made available for the entire current solar cycle. These compare favorably with determinations by other stations using the same scale. Meaningful daily indices using these data are also available.

Indirect Detection of Flares—Instrumentation

Objectives: For this project at the High Altitude Observatory three automatic recording instruments for indirect flare detection utilizing enhancements of atmospherics and absorption of galactic radio noise were designed and constructed. These were provided for use at Boulder, Sacramento Peak, and the McMath-Hulbert Observatory; from Department of Army funds, a fourth instrument at Hawaii was added to the net in late IGY. These instruments contributed to the international solar patrol operated for the benefit of the upper atmosphere research program.

Results: The data were used for the warning network and also compiled and published in the High Altitude Observatory reports of solar activity and the NBS F-Series, Part B, "Solar Geophysical Data." From the operation of one of the units at Boulder, it was discovered that the SCNA recorder was sufficiently stable to permit quantitative measurements of the absolute change in transmission of the signal through the ionosphere. From this, project operators were able to calculate the variations of ionizing flux radiated by flares.

McMath-Hulbert Flare Patrol

Objectives: The purpose of the project at the McMath-Hulbert Observatory, University of Michigan, was to carry out a continuous patrol of the sun, whenever weather permitted, during which motion picture records were obtained of solar activity in the light of the red hydrogen line. Existing equipment at the McMath-Hulbert Observatory provided only for high magnification studies of relatively small areas of the solar disk, and equipment was acquired for the continuous recording of solar flares over the entire disk. The observations and measurements were carried out at the McMath-Hulbert Observatory where extensive auxiliary facilities were already available in the form of motion-picture processing equipment, microdensitometers, and other measuring devices. The services of several members of the Observatory staff were contributed without additional cost to the project.

Results: During the ten months that the flare patrol was in operation (1958 March to December), the telescope was in use for 1,391 hours. In this interval 452 events classified as flares and 1,153 events classified as subflares were recorded.

Solar Radio Noise Patrol—Ithaca

Objectives: This 200 Mc radio noise patrol at Cornell University was operated as a part of the IGY program as an adjunct to a larger program at the University in radio astronomy. The Office of Naval Research supported this program for some time, but IGY funds were required to continue operation of the solar radio noise patrol from mid-April through December 1958.

Results: The reports of large solar disturbances aided the IGY World Warning Agency (AGIWARN) in its effort to forecast geomagnetic disturbances. The flux data itself forms a part of the complete solar history of the period.

Flare Patrol Equipment

Objectives: This project at the High Altitude Observatory of the University of Colorado provided support for optical flare patrol instrumentations at the University of Hawaii, Sacramento Peak, and the High Altitude Observatory.

Results: The three flare patrol stations (Sacramento Peak Observatory, Climax, and Hawaii) operated with a high degree of efficiency during the IGY period, affording essentially complete patrol coverage during the daylight hours between midwestern United States and Hawaii. Reports of flares and also prominences were submitted by all three observatories to the World Data Centers and their data was also incorporated in various intermediate solar publications.

Solar Activity Data Reduction and Publication

Objectives: This project at the Central Radio Propagation Laboratory (CRPL), Boulder Laboratories, National Bureau of Standards, provided for expansion of the format of the solar-geophysical data published by CRPL on an intermediate time schedule as well as for reduction of new

U.S. solar activity data and for the preparation of selected manuscripts, requested by CSAGI, for publication.

Results: Research workers throughout the world received these solar-geophysical data compilations which allowed them to prepare prompt evaluations of their projects. The solar flare data in the WDC-A: Solar Activity were placed on punch cards and were continually revised. These cards were used in a high-speed electronic computer program to combine the multiple reports of solar flares into a single, statistically weighted event. A Calendar Record for the IGY of selected solar and geophysical events was prepared. It was submitted to CSAGI for publication in the *Annals of the IGY* by A. H. Shapley as part of the World Days and Communications Program.

Interview with Frank Niepold, MSEd

Senior Climate Education and Workforce Program Manager
and Coordinator at National Oceanic & Atmospheric
Administration (NOAA)

Mark O'Connell: Can you tell me about your work as a climate educator?

Frank Niepold: Let me give you a little bit of context of who I am and how my answers sit in here a little bit differently than most. I had a really unique opportunity because I was in Prince George's County, teaching science to middle-school students. And a lot of the NASA Goddard Space Flight Center scientists, their kids went to the school. And one of them happened to be the head of the program that had the Landsat mission, the longest satellite observing Earth mission ever. And he asked, would you like to go to Saskatchewan and do some research with us, as part of your teacher professional development? I said, yes. So that was a really cool opportunity. I learned a lot. But what was happening in the scientific field was—this was 1994—the earth science community really started thinking they had enough data of the earth to really understand each piece of it, and how all those pieces came together to really understand the earth, and within that the climate system. And so they passed a law called the Global Change Act of 1990. And it really drove an earth system science transformation in NASA Science Foundation, Department of Energy, NOAA, and the White House. And out of that a whole host of things came together, whole new fleets of satellites came in, supermodels came in.

But the transformation from earth science to earth system science was still happening at NASA. And I was part of something called Mission to Planet Earth. A great mission, like what are all the kinds of satellites that we need to put in space to see all the processes of the earth at a planetary level to protect the home planet? So I was at NASA during that time, and I was like a kid in a candy store, man, it was amazing. The best scientists in the world and I'm just learning of system science from the best in the world.

At the time, climate wasn't really very comfortable to talk about because this was under George dub, you know, under W. And yet polar research under the International Polar Year (IPY), because it's an international thing, really grabbed the community and allowed us to focus on how to build our work together with a focus on IPY, but really, it was the foundation of really starting to focus on climate change. Because we were studying the planetary system: Why is it changing? What are we seeing? How has it changed from the previous IPY? Why the relationship?

But then there's a group of us who are like, well, I get the polar part. But what about the nonpolar part? And so there was this thing about the poles and then the teleconnections between the poles, which gets you climate change. And so we were starting to kind of inch our way into it. But I came from this program where part of my job at NASA was to connect system science to what students and scientists and teachers could do with relatively inexpensive materials spread all over the place using common protocols shared. That is, in essence, the same spirit of the original IGY. Because what was going on at the time was people were going out and taking a measurement in a place, and they were using a certain methodology and a certain place and they weren't sharing. And so doing the same measurement the same way in different places and sharing at the same time really opened up a really kind of transformational moment in the field. Because it was like, if we don't do it the same way at the same time, and looking at the same thing, and don't share, you get the same problem.

But it was even worse than that, because they were working in different ways in different places, at different times, and not sharing before that. So it was just a really interesting story that opened up the idea of, can you understand a planet? And the answer was, you could understand a part of the planet, which is IPY. And so that really was a seminal moment in humanity, a really important transformation in our understanding. The same is true for satellites. The same is true for ocean buoy arrays. The same is true for weather balloons, where you're measuring, you know, temperature and pressure going up with a weather balloon twice a day, and all these places over and over again, and sharing the results. That's how you understand planetary process.

M.O.: Okay, so you're at NASA, and you end up at NOAA? How did that transition take place?

F.N.: There's a story. In 2002–2003, the moon and Mars missions started showing up at NASA. And NASA loves a presidential priority. They just love them. And they're like, yeah, let's go do that, and then have this, oh, let's go do that. And the zigzag that happens as a mission is substantial.

So I was working on Mission to Planet Earth. And I'm a very mission-oriented person. You give me a great mission, and then let me get at it, and I'll do amazing things for you. But don't change the mission on me halfway through. So I was still on it. I was solid. I was all in. And they started talking to us about, well, let's really look at the priority of studying other planets. But I'm like, we're pretty much on home planet, because that's everything that we love. It provides this life support for everything we love.

But we don't totally have a good understanding. I mean, this seems to be way more important than the moon or Mars. And we would have these battles inside the agency. And they're saying, no, no, no, move to Mars. And I'm saying, what, are you crazy? I mean, this just doesn't work. So I lost that battle. But so then I thought, whoa, who doesn't lose track of a home planet? Because it's kind of the most important thing. And an opportunity at NOAA showed up. And the opportunity was to become the first Climate Education fellow. And the charge they gave me was to make the nation climate literate. And I thought, now that's a mission. That was eighteen and a half years ago. So I'm still on it.

M.O.: I would love to understand what "climate literacy" means.

F.N.: In 2005 the ocean community had realized that the oceans were not properly being taught, focused on in any form of education. So they created something called the *Ocean Literacy Guide*.

So NOAA started the Climate Education fellowship, and when I got selected, I was charged to go forth and do the same, and here's the Ocean Literacy project as a guide. And I realized that y'all had a whole community (of oceanographers), so I guess I also have to start with community and partners. And so that's what I did. We built a whole coalition of partners and organizations and networks and structures, and we did the first version (of the guide) in 2008. And it has been used domestically, it's been used internationally. And really what the *Climate*

Literacy Guide asks is, what should everybody know about climate change? You should understand how to discern credible information about climate and should be able to make informed and responsible decisions about climate. And then the guide's ideas were really about how energy works in a variable system, how humans are impacting it, how are we changing it? And what are the impacts that are affecting us and ecosystems? It was used by the National Academies to make the next framework for science education in the United States, which now forty-eight states use, and it went from like a really niche project topic in science education across K–12, to about 30 percent. So that's a massive shift. And now that wave of effort is . . . moving beyond science to include all the disciplines. It was used in museums, zoos, aquaria, domestically, internationally, it's a big deal.

And this guide was so there was an engine and now the blueprint was there for knowing what to do with the engine, and it was starting to be used. For instance, right now it is the policy of the U.S. government that every agency has to move forward and advance climate priorities. And they're also required to have a plan, and in that plan is how are you going to educate your workforce, which means labor, treasury, transportation, defense, NASA, NOAA. And you know, this is being used to inform that work. Every job is going to be a climate job, which means everyone's going to have to know enough to be able to use climate in their information. So climate literacy is about to explode. Because it's now going to move into the adult dimensions, as well as the young people dimension. We're talking about colleges and universities, businesses, industries.

If you're in the agriculture sector, if you're in the insurance sector, if you're in the energy sector, if you're in the food sector, I mean, think of a sector of society, and then think, how does that not have a direct relationship to climate in one way or another? Either you're contributing to making climate change worse, in your business sector, or you are a part of the solution to climate, or you're trying to remove the vulnerabilities to climate [of] your work. So it's about to just go through the roof.

M.O.: I found an interview, posted by NOAA online, conducted with you. There's one quote from that interview that I just loved, and they even used it as a callout quote. You said, "Partnerships. That's where the work gets done." So naturally, I'm very interested in hearing your thoughts on that.

F.N.: It's still true. So absolutely true. Because, when they said, Frank will make the nation climate literate, what was the first thing I did? I had no staff. I had no budget. I had no frameworks. I had no partnerships. I had nothing. And so then I was like, well, my first move, because I can't have staff and budget yet, was to seek out partnerships. And so there's an organization called Project 2016, one that was being done by the American Society for the Advancement of Science. And I needed to talk to them, but I couldn't get their attention. Because I'm just, you know, I'm just this new climate education fellow. So I went down there, I got to their front door and I rang the doorbell. And the guy that steps out is Ted Willard. Ted Willard is a really big deal in this space. And he just pokes his head out, like, yes, what can I do for you? And I said, Ted, I'd like to talk to you about this crazy thing. And he goes, it's funny you asked me that, because I'm actually just starting to work on the weather and climate part of this. Could we partner? And I'm like, absolutely.

It's been like that for eighteen and a half years, where everything that's being done is done through partnerships, whether I'm partnering with the White House, partnering with international agencies, partnering with state governments, universities, museums, zoos, aquaria, NGOs, you name it. If I hear about a new initiative, I just call [that office] up as soon as I learn about it, and they're like, how did you find out that this happened? Well, because I'm a really good digger. Because I need all the partnerships, because it's always going to be a "we" scenario. I mean, it's just how it works.

So what's interesting about climate is it forces everyone into a collaborative posture. And I think it's because it's so overwhelming that no one feels like, Oh, I've got it. We're good to go. I don't need any help from anybody. That is not true at all. And so every time I reach out, I'm almost always greeted with, yeah, sure, let's talk, let's see what we can do together. And, you know, that's been the recipe for my success. I tried to do it when I was at NASA, and everybody was competing. When we were working on remote sensing education and ERP (Enterprise Resource Planning) system education, everybody's competing. As soon as you got the climate, it's like, a posture changed to a very collaborative position. And it's been wonderful.

M.O.: So it is feasible. It's feasible to think that maybe the magic formula IGY from 1957 could be replicated in the scientific world now. Do you think there's potential there?

F.N.: Well, I would even go further, because the scientific world climate has shifted. I was part of the U.S. delegation to the Paris Agreement, where people from 196 countries came together and said: Could we actually figure out how to solve climate change together?

Is there still time to lessen the impact? Could we actually do something to undo some of this damage? And that, for me, is really what most of the work right now is dealing with. Can you decarbonize your electricity? Well, this is a common problem for 197 countries. Can you decarbonize your transportation systems, can you decarbonize your food systems? Can you decarbonize your forests, can you figure all these things out? Can you reforest your deforested lands?

So what I'm trying to get at is the big thing. Doing the same thing in different places, in the same way, and sharing the results, for me, is about solving climate change. So for me the amount of energy that young people want to put into studying the problem is really diminishing. What they really want to get on is solving it.

I ask students all the time about this, and I say, look, are you being taught climate change? If you are, are you happy with how it's being done? How much problem versus solution do you want to focus on in your studies? Once they understand the question, they'll say, "I want to understand the problem, but I really want to understand how we can build skill and knowledge to solve it. That's really what I want." And I love that answer.

M.O.: Are there other ways in which you see partnerships and collaborations being formed in earth sciences?

F.N.: Any data we caught, or we produce, whether it's a model output that is global, a satellite product that is global, is available to the world for free. Most countries don't do this. Most of them protect their data from their own citizens and/or other countries. All of our data is free. So when I used to work for Landsat, part of my job was to help people understand how to access, download, manipulate, do analysis on, and use Landsat data from anywhere in the world. And other countries were just gobbling up our data for their own national or domestic purpose. That's just that same spirit of access, free access for the benefit of all, still operating in our methodology today.

But just that idea of sharing, and similar methodology to really understand we need to understand this together, if we do the same thing the same way, with similar techniques, at the same time, and share that in different places. If you do that, we actually might be able to understand a very large part of planet Earth. And that turned out to be true.

M.O.: What do you consider your greatest accomplishment, in terms of learning about the earth, or fixing the earth?

F.N.: I didn't accomplish anything of any merit. But because I worked with some amazing people, *we* did. And you know, we've proven that education is incredibly important in how we respond to the climate crisis. We've done that. We have found that education is an incredible foundation to building the workforce that is needed to solve climate change. We did that. We've established both of those things. We've proven that all forms of education are important, not just to classrooms or museums, or aquaria, or colleges, or career technical centers or community education, all forms. And we have heard from young people that they would like a different education than they were seeing, and they're willing to work with us to figure that out together. And the last one I would say is that we've heard strongly from indigenous communities, that they have a lot to teach us. And they're willing to extend that offer, still, in spite of everything, and that is a great humbling opportunity for all of us, that we could listen to that way of knowing and still have a lot to learn from them.

So even with as much as we've accomplished, we also have a lot more to learn in this way. And so I think that my greatest accomplishment is that I was asked to stay on something and just keep on it. And I did, and in the doing of that, we have done something of significance. I will say that I still see sometimes people try and make it about themselves and they try and compete, and it just doesn't work. And they find themselves by themselves often because the collaborators like to work with people who collaborate, and they like to partner. So I still see we've got a good bit of work in front of us, but you know, these are important times, so it's a great honor to be able to have this role.

CHAPTER 4

On Ice

Now we travel south, to the land of penguins, icebergs, and the driest air you've ever taken into your lungs. Before I get too far into the role of Antarctica in IGY, however, a personal note: My late father, John "Gene" O'Connell, was a world traveler and was very proud of the fact that in his lifetime he had traveled to six of the seven continents. But six wasn't enough. My dad, a guy who had lived well into his nineties shunning excess and being content with the simplest of pleasures, harbored a burning desire to go a little nutty with the checkbook and set foot on the seventh continent, Antarctica, to make his globetrotting continent-hopping complete. So in 2015 he invited me and my siblings to join him on a two-week Antarctic cruise. My mom at that point was living in a memory care facility and couldn't travel (though if she had been able to travel, she still would have passed on the trip; icebergs and penguins were not her cup of yerba mate).

The whole trip was an exercise in adjusting one's sense of scale. There is a vastness to Antarctica that defies description, and the word "cold" loses all meaning. Not a day went by that I didn't shout out, "You have *got* to be kidding me!" or "No way!" or exclamations to that effect while beholding a scene of unimaginable Antarctic grandeur. The effect was only magnified when I would look back at our cruise ship from land and would see how ridiculously dwarfish it appeared when set against the snowcapped peaks of the continent. It makes you feel shockingly small. We all read Shackleton before leaving on the trip to prepare ourselves for the barren icescape, and we were still humbled and unprepared.

All of which is to say that of all the chapters of the grand tale of IGY, the research done on the seventh continent hits closest to home for me. I may have only stood on the continent for what amounts to just a few brief moments, but I have an appreciation for what it feels like to be there, to see it and touch it and let its chill work its way through my parka. I still feel a thrill of cold rapture thinking about it. That said, it almost makes me laugh to think that as recently as 1957 Antarctica was considered a land of deepest mystery and danger. I mean, '57 wasn't *that* long ago, was it?

Not long ago I had the pleasure of interviewing Louise Huffman, MSEd, director of education and outreach for the U.S. Ice Drilling Program, and found her Antarctic journey to be every bit as wondrous and awe-inspiring as my own. "I think that thing that gets people so excited about it is it's like being on a different planet," Huffman told me. "It's otherworldly, it's exciting, it's dangerous, it's beautiful, but in a way that's just totally out of this world."

While I had the luxury of traveling to the Antarctic in a warm, comfortable cruise ship that was never more than a few minutes away, Huffman lived in a tent for three very chilly months. "When you're there—you probably experienced this, too—when I stepped out of the plane on the ice shelf at McMurdo Station, the air was so crisp, and cold and clean, it's the cleanest place in the world," Huffman went on. "And when we were out in the field, there were so many ways that you just felt on overload. Because your senses are on overload. You're seeing these great expanses of white in the dry valleys, great expanses of rock. . . . But you also can see forever because it's flat, and most of the places where scientists work there's nothing there to give you any idea of distance or size. When we were in the field, and we'd look at the mountains we'd have no idea how long it would take to walk over to that mountain, it looks like it's right there. But there's nothing there to give you any idea of the landmarks. But your senses are also deprived in a way that's weird. You don't hear anything, the wind doesn't make any noise. It's very windy, but there's nothing, there're no trees to make noise. There's no rustling of leaves, there's no background traffic noise, so you don't hear anything. Unless the wind's flapping your tent."

"There are no smells," she enthused. "There's hardly any bacteria at all. So you know, we can keep our food outside, there're no animals that are gonna need it. There are no bugs. There's no smell (unless you happen to be standing right next to the latrine). But other than that, there's no smell, and no color. So, your senses are just kind of wacky, wacky. Yeah."

Wacky, indeed.

You'll remember that twice before, the world's scientific leaders pooled their resources and talents to explore the most mysterious corners of the earth, and both of those events were known as International Polar Years. In a way, that makes IGY the youngest sibling of our three "Years" and one worthy of some special attention.*

Up until IGY, only a few thousand humans had spent much time living and working on Antarctica, and so no one country could claim it as its own. Thirty-one bases were built, either on the continent or on the ice shelves, by a dozen countries. Seven of those countries considered certain swatches of territory to be under their watch, but the claims were symbolic, almost fanciful, as the extreme geographical and climatological conditions had until IGY discouraged thoughts of large-scale, long-term conquest. To drive that point home, some chroniclers of IGY used combative terms like "assault," "attack," and "onslaught" to describe plans to explore the mysterious interior of the continent. "It is the land of the absolute," wrote J. Tuzo Wilson. "No one comes here casually."[1]

---

* Because the IGY was originally conceived as an International Polar Year (IPY) before its scope was officially broadened to include the whole of Earth, the sun, and space, some people quoted throughout this book use the terms IGY and IPY interchangeably. In cases where a quote refers to the event as IPY, I have (with their approval, where possible) secured the person's permission to clarify the references. There was also an International Polar Year (also IPY) in 2007–2008.

By the time Wilson made his trek to Antarctica in the company of Lloyd Berkner, IGY work was already well underway, bustling with what *Life* described as "a concert of the world's scientists and armed forces, mobilized in an unprecedented example of nations' cooperation, to unveil the land that had lain in forgotten, frozen slumber."[2]

The U.S. IGY project in Antarctica had begun in 1953 with the establishment of six bases around the continent, at which government agencies such as the U.S. Hydrographic office and the U.S. Geological Survey would conduct research alongside private institutions such as the Scripps Institute of Oceanography and Cornell University. Logistics were to be handled by the Department of Defense, which had a great deal of recent experience coordinating the delivery, construction, and use of huge amounts of supplies and equipment in remote areas. Research at the bases would encompass meteorology, ionospheric physics, aurora, geomagnetism, and cosmic rays. Four working groups were formed early on to consider Antarctic policies, logistics, personnel selection, and perhaps most important of all, "the type and quantity of cold weather clothing required to outfit the U.S. IGY Antarctic scientists."[3] Can't forget the anoraks when you're heading to a place where winds can blow up to two hundred miles per hour and temperatures can reach 125 degrees below zero F.

The six bases established by the United States were led by the quaintly named Little America Station, built on the Ross Ice Shelf. In addition to coordinating activities at the other five stations, the staff of Little America were in charge of the International Antarctic Weather Central and its associated international communications network.

If you were looking for an IGY gig that promised true isolation, the Amundsen-Scott South Pole Station was for you. Situated a mere thousand yards from the magnetic South Pole, the staff of this very remote base specialized in glacial research, studying the effects of ionospheric activity and auroral phenomena.

Byrd Station, an inland base, also concentrated on studying the aurora and was additionally tasked with studying "pressure surges" in the ice (sudden eruptions of ice) that were first observed by the Scott Expedition fifty years earlier.

Ellsworth Station was built on the Filchner Ice Shelf in the Weddell Sea. Along with Little America and Byrd Station, Ellsworth served as a launching point for several traverses into and across the continent by airplane and surface vehicle. Studies of "Whistler" waves (very-low-frequency electromagnetic waves) amid the earth's radiation belts were conducted here with the close cooperation of the British and Argentinian teams.

Hallett Station was a center for meteorological science, conducted with the cooperation of the New Zealand team. Scientists at Hallett studied the gamut of geophysical phenomena, and the base maintained a crucial emergency airstrip for airplanes flying between New Zealand and McMurdo Sound.

Wilkes Station was built on bedrock, which made it a perfect site for seismological studies. Wilkes scientists also took part in the study of cosmic rays, oceanography, and surprisingly enough for a continent with few resident animals, zoology.

McMurdo Naval Air Facility on Ross Island in McMurdo Sound was (and still is) the base of operations for air travel between New Zealand and Antarctica, as well as a supply warehouse and distribution center for the far-flung U.S. network of bases. It was also home for a decade to a small nuclear power plant provided by the U.S. Navy. Although the device was successful in reducing the use of petroleum fuel to heat and light the base and could be used to augment some scientific undertakings, hairline cracks and water leaks convinced the navy to prudently decommission the reactor in 1972.

The bases formed a cozy sort of polar community where countries that might never have interacted before IGY now found themselves neighbors. Where else in the world would you find Australia elbow to elbow with France, or Chile overlapping the United Kingdom? The exigencies of Antarctic living rendered national borders irrelevant and led to previously unheard-of collaborations. If a member of one country's crew was injured or fell ill, a helicopter or plane could arrive from any country's base to airlift the injured party to safety. If an expedition ran into difficulties, crews from the nearest base (no matter which country's) would show up with needed supplies or equipment, no questions asked. Everyone was responsible for everyone else, and it seemed the most natural way to be. It's difficult to imagine how any of the Antarctic teams could have succeeded in their missions if it hadn't been for the generosity and goodwill of their neighbors.

*   *   *

"Why all this interest in this wasteland?" asked the writers of *Life* magazine.

"Remote as it is, Antarctica has a highly strategic location," the magazine replied to its own question. "The continent could be a key to military control of the southern hemisphere by planes, ships and missiles. The Navy considers the Drake passage (the body of water that separates South America from Antarctica) essential for deployment in case anything happens to the Panama Canal."[4] If that portentous declaration comes as a surprise, you may be equally shocked that in the optimistic mood of IGY, major commercial airlines were excitedly pondering the economic feasibility of transpolar passenger flights connecting the southernmost reaches of South America, Africa, and Australia.

The benefits of Antarctic research to science were expected to be immeasurable, as *Life* pointed out: "For scientists studying the Earth's past, Antarctica is a rich untapped lode of knowledge. From studying the layers in long cores of ice, reamed from the continent's thick covering, geologists can tell the annual snowfall for the last thousand years."[5] Bubbles of gas trapped under many layers of ice, it turns out, have an interesting tale to tell—if they can be retrieved intact.

Reading those ice-borne stories from the past can tell a scientist a lot about the future. During IGY, scientists at Little America found that they were in the midst of a small but significant thaw. They discovered that in the previous fifty years the temperature at the site of the base had risen five degrees, and they calculated that if the temperature continued to rise at that same rate, all the ice on Earth would someday melt, and global sea levels would rise by two hundred feet. It's interesting to think that in the time of IGY, more than sixty years ago, global warming was attracting at least a moment of attention from the Little America scientists and the editors and readers of *Life*. But here's the catch: no scientists at the time of IGY believed that a small thaw was of any concern or ever would be. More on that subject later.

*   *   *

About those ice cores: When I started my research for this chapter, I began to get the distinct impression that scientists—both our own and those from other countries—were drilling these super-deep holes willy-nilly all over the continent, creating a Swiss cheese effect that must be visible from pretty much anywhere you stood on the continent (not to mention causing you to be careful where you step). So, the first chance I got, I addressed my concerns to Huffman: "I know you and your team do a lot of drilling for ice cores, as do, I'm sure, all the other countries, scientists, and engineers scattered around Antarctica. So, there must be thousands of drill holes all over the continent. Hundreds of thousands of drill holes through all these decades of drilling. What have we learned from those thousands of cores? What have we learned from all this drilling?"

"Well," she corrected me, "first of all, there's not thousands of holes, because to get an ice core you often will take a couple of years to go down just one hole. So the deepest ice core was about two miles, and it was near Vostok Station (Russia). But the different ice cores are usually on domes, the domes of ice because they want the deepest ice possible that they can get."

"And, honestly, ice cores tell us more about climate history than any other science," Huffman added. "It's probably the most important science when talking about climate change. It's absolutely vital that we keep studying and keep doing these ice cores in Greenland and

Antarctica because of the things that we're learning. But probably the most important thing that we've learned is that we have an 850,000-year record, continuous record of both carbon dioxide and temperature. This is a direct measurement right out of the bubbles of the ice cores. We actually have direct measurements back almost a million years in our climate, and what we've learned is really amazing."

*   *   *

A glaciologist who embarked on not one but two oversnow traverses was geophysicist Charles Bentley. His work of measuring and interpreting ice core samples to determine the structure of West Antarctica could only be done as part of a thirty-day, 640-mile traverse from Little America on the coast to Byrd Station in the interior. "We collect our vehicles; 3 Tucker Sno-Cats," Bentley later recalled. "Gasoline powered, 4-track drive, all pontoons turn, cab over engine for warm access to latter—also keeps cab warm—so warm, in fact, that often we have to drive with open windows and doors to cool off. Benches inside for instruments and sleeping, cooking, and scientific equipment. Aneroid altimeters, exploration seismic gear, gravity meter, magnetometer, rammsonde, snow density gear, radios for inter-Cat and traverse-to-base communication, navigational equipment. Traverse routine: travel one Cat ahead then other two five miles behind for differential altimetry. We stopped every five miles for altimeter, gravity, magnetic readings, rammsonde (measuring the penetration hardness of snow) every 20 miles; alternate days traveling, then day on site for seismic work and pit studies."[6] If it sounds like a tremendous amount of work punctuated by long stretches of tedium, it was. Dangerous as well, as the crevasse detection equipment didn't always function as intended, so gaping pits could open up underneath the Sno-Cats at any time. Then there were the inevitable fights over who should prepare the meals (Bentley made breakfast every day, as he was an early riser; the other meals had to be negotiated). Despite these hardships, the traverse was a smashing success.

Bentley grew far more eloquent some time later, when he related his experiences working on IGY as part of an oral history interview conducted by the American Institute of Physics (AIP) in 2008.

According to what he told the AIP interviewer, Bentley happened upon his icy career somewhat randomly, after a school aptitude test steered him toward physics. After earning his undergraduate degree, he spent a few years working as a marine geophysicist, until one day in the early 1950s a colleague asked whether he'd like to go to Antarctica. "I thought that sounded like a pretty good deal," recalled the agreeable Bentley. The colleague turned out to be a recruiter for IGY, and before he knew it, Bentley was conducting geophysics research in Greenland, learning how to work on an ice shelf in preparation for his season in Antarctica. That season turned into a twenty-five-month marathon and a life's work.

"What happened was that I only was expecting to go for one year. But I got so interested in what we were finding, which was how deep the West Antarctic ice sheet is, how far down the bed is, that I—believing that the IGY was a one-shot deal, that it was going to be over in another year, and I wasn't through finding out what I wanted to find out—I re-upped, because I wanted to complete the survey of the area around Byrd Station in Central West Antarctica."[7]

"Other national expeditions, particularly in those days, routinely sent people down for a couple of years," Bentley went on. "You could only go in by ship, and you could only get in by ship at the end of the summer season when the sea ice had retreated. So, in order to get a full summer's field season in you had to go in the end of the field season before and spend the winter first, and then do the field season, and then come home or not."[8]

Bentley chose "or not." But he got bored and lonely and found himself stateside a year later, visiting New York City and experiencing human contact in a whole new way. "One of the first things I did was make a point of getting on the subway at rush hour. Because I had seen so few people for over two years that I wanted to—I just enjoyed the crush, which I remembered. I mean, I had been on it before. I never really appreciated the New York subway at rush hour back when I was living in New York, when I was at Columbia. But, after two years in the Antarctic I had a different feeling about it."[9]

There are reasons that I find Dr. Bentley a fascinating figure. What kind of person, I wondered, would stay in lonely, cold Antarctica a year longer than he had signed on for because he wanted to finish measuring the thickness of the West Antarctic ice shelf, and who would intentionally board a packed subway car to savor the feeling of crushing humanity? I'm also fascinated by Bentley because of a personal connection. I have a friend named Jim who lives in Wisconsin but works in Antarctica in the summer season "digging holes for the science guys," as he describes it. When I asked Jim about his Antarctic experiences, he told me about the late Dr. Bentley and directed me to Bentley's works at the University of Wisconsin–Madison, where he founded the geophysics program. Jim told me he had worked with Bentley and considered him the preeminent scholar of Antarctic science. By pure chance, I had just finished reading Bentley's oral history, which seemed somewhat synchronistic, and here we are, honoring a guy who enjoys being jostled on the New York subway.

Here, in his oral history, Bentley described in detail how he made his first IGY discovery while on a traverse from Little America Station to Byrd Station in February 1957:

> The trail out to Byrd Station had only just been finished a few weeks before and they were just starting to put up the station, because they had a lot of trouble getting through the crevasse zone, the boundary between the Ross Ice Shelf and inland ice. But we had a marked trail, so we drove our vehicles out from Little America to Byrd as that was a way to get the vehicles there and also to do some work along the way.
>
> So, we were doing seismic sounding and other geophysical and glaciological work along the way. And, that was when we first discovered that the bed of West Antarctica was so far below sea level. Up until then we had expected that since there were mountains to the north of the route

from Little America to Byrd, and there are mountains to the south that there were probably mountains covered by the ice in Central West Antarctica also. So, we were quite surprised—we thought after we left the ice shelf where it's floating, of course the bed was well below sea level there, it has to be, but we thought when we got to the grounded ice that the bed would come up underneath. But instead, it went down.

And, at first I wasn't even sure that I was making measurements right.[10]

But Bentley's measurements were correct, much to his surprise. It seems that because he expected the bed to be rising above sea level, he was blind to the possibility that it might actually be slanting down below sea level.

Bentley's soundings seem straightforward and foolproof enough: simply measuring a signal creates a signal that is bounced between the ice surface and the bed. A single reflection is a signal that bounces off the bed and is measured again at the surface. A double reflection bounces off the bed, the surface, and the bed again, giving a measurement at the surface of apparently twice the depth of a single seismographic measurement. "This is what I thought it was," he said. "That's what it appeared to be. But this bed was so deep that I didn't really believe it."

"So, it was in the course of that February drive out to Byrd," he recalled, "that we discovered that the West Antarctic ice sheet had a bed way below sea level, which is why it's been in the news . . . because of potential instability." Finally accepting the truth of his seismograph soundings, Bentley continued on with his team to Byrd Station, continuing to take readings along the route. "Well, it just kept going down. Byrd Station is quite a long way inland. And, it just kept going down as we went into Byrd Station, and then, of course, I was really anxious to see what happened when we went north and east."[11]

What he saw happening to the north and the east is that the bed continued to descend deeper and deeper below sea levels, defying every assumption ever made about the region, and opening up a whole new field of research for future expeditions. It is moments like this that define the spirit and intent of IGY.

Bentley's further experiences as he wintered at Byrd Station took him on three-month traverses around the bases. On these voyages, the scientists would live in their Sno-Cat treaded ice vehicles. The basic crews were two seismologists, two glaciologists, a mechanic, and the occasional adventure-seeking hitchhiker from another base.

"So, then we had a regular routine of traveling a day and, and then doing station work for a day," Bentley went on. "We would travel twenty-four or thirty nautical miles one day and then the next day we'd do our seismic sounding and detail study of the wave velocities in the upper part of the ice sheet. And, the glaciologist would dig a pit to study the stratigraphy within the snow layers so they could determine what the snow accumulation rate was, primarily. That was a day's work, and then the next day we'd travel again. And, while we were traveling we stopped every three miles for gravity measurements for the geophysicists, and for snow hardness measurements for the glaciologists. And then, we were doing altimetry also, aneroid altimetry. And, the way we got it, tried to get rid of weather effects, was by traveling three miles apart, so we were always reading two altimeters over a three-mile interval. Then we'd assume that the large-scale pressure changes were essentially the same at both ends of the three-mile line. So, the differences in what the altimeters read were the elevation differences between the two spots. So, then we got the absolute elevation just by adding differences, which is an inaccurate way of doing things. You know how errors can accumulate. But, it was the best we could do because . . . it was better than just having a single altimeter. Well, then, you're measuring nothing but the weather."[12]

Back in the United States, Bentley spent the next few years working with his IGY data and then eventually developed the Antarctica research program at the University of Wisconsin and continued his analysis of Antarctic ice core samples. "Everything you read about rapid climate change—the evidence for the climate can change over the course of just a few decades—that all comes from ice core information," Bentley said. "And [evidence of] past changes in temperatures and carbon dioxide levels in the atmosphere, that all comes from ice cores. All of that is deep ice core stuff.

"I think there was general recognition fairly early that ice sheets could change rapidly enough to cause significant sea level change on short time scales, speaking from a geological perspective. Where the short geological time scale is quite different from a short human. Well, I mean, it turns out that in some cases that's not true. I mean, significant sea level rise within human lifetimes. But there was a lot of skepticism about how fast it could happen, but I think it was recognized as something that really could and did happen. It's all a matter of how fast."[13]

"I think the visibility of the IGY Program in the Antarctic was the beginning of recognition that glaciers and ice sheets really are an important part of the world system," Bentley concluded. "And that has just grown as people have learned more and more about how ice sheets and glaciers have changed through time, and how much they interact with the environment, with the climate, with the ocean levels. But, yeah, I think you can say IGY was a turning point as far as increasing the activity and the visibility and popularity of glaciology."[14]

So important was the IGY work performed in Antarctica that it spurred the development of what may be the most remarkable and resilient multinational agreement in human history. Where most treaties are marked by compromise, with all parties involved conceding one point or another and accepting at least some of the terms desired by the other party or parties, the Antarctic Treaty was defined by its unanimity. No one gave up anything to reach this agreement, because in a sense the treaty had already been operational for eighteen months due to the work that was already being done under IGY.

The Conference on Antarctica was initiated by the United States in the waning months of IGY, attended by representatives of the twelve countries that had cooperated on IGY research in Antarctica: Argentina, Australia, Belgium, Chile, the French Republic, Japan, New Zealand, Norway, the Union of South Africa, the Union of

Soviet Socialist Republics, the United Kingdom of Great Britain and Northern Ireland, and the United States of America. "As a result of the deliberations of the Conference, as recorded in the summary records and reports of the respective Committees and of the Plenary Sessions, the Conference formulated and submitted for signature on December 1, 1959, the Antarctic Treaty."[15]

What was referred to as The Final Act of the Conference on Antarctica was then written up in English, French, Russian, and Spanish; signed by all participants; and deposited in the archives of the U.S. government. In essence, the document "sets Antarctica aside as a preserve for scientific research."[16]

Here's what this remarkable document says:

> Recognizing that it is in the interest of all mankind that Antarctica shall continue for ever to be used exclusively for peaceful purposes and shall not become the scene or object of international discord; Acknowledging the substantial contributions to scientific knowledge resulting from international cooperation in scientific investigation in Antarctica; Convinced that the establishment of a firm foundation for the continuation and development of such cooperation on the basis of freedom of scientific investigation in Antarctica as applied during the International Geophysical Year accords with the interests of science and the progress of all mankind; Convinced also that a treaty ensuring the use of Antarctica for peaceful purposes only and the continuance of international harmony in Antarctica will further the purposes and principles embodied in the Charter of the United Nations; have agreed as follows:

> **ARTICLE I**
> 1. Antarctica shall be used for peaceful purposes only. There shall be prohibited, inter alia, any measures of a military nature, such as the establishment of military bases and fortifications, the carrying out of military maneuvers, as well as the testing of any type of weapons.

2. The present Treaty shall not prevent the use of military personnel or equipment for scientific research or for any other peaceful purpose.

ARTICLE II

Freedom of scientific investigation in Antarctica and cooperation toward that end, as applied during the International Geophysical Year, shall continue, subject to the provisions of the present Treaty.

ARTICLE III

1. In order to promote international cooperation in scientific investigation in Antarctica, as provided for in Article II of the present Treaty, the Contracting Parties agree that, to the greatest extent feasible and practicable:

    (a) information regarding plans for scientific programs in Antarctica shall be exchanged to permit maximum economy and efficiency of operations;
    (b) scientific personnel shall be exchanged in Antarctica between expeditions and stations;
    (c) scientific observations and results from Antarctica shall be exchanged and made freely available.

2. In implementing this Article, every encouragement shall be given to the establishment of cooperative working relations with those Specialized Agencies of the United Nations and other international organizations having a scientific or technical interest in Antarctica.[17]

The treaty goes on for an additional eleven articles, addressing such issues as geographical boundaries of the treaty; schedules of ongoing future administrative meetings in Canberra, Australia's capital; and methods of solving any future disputes, should they arise. Signatory nations saw the treaty as a way to make Antarctica "immune from military and political strife."[18]

A further article in the treaty granted the member nations the right to name observers who would be free to visit and inspect the programs of any other signatory country at any time and for any reason. Author Sullivan called it a "free-wheeling inspection system" that "revived the principle of unanimity,"[19] and *Life* magazine noted that this was "the first inspection system ever agreed to by both the U.S. and the U.S.S.R."[20]

This leads us directly to Article V, undoubtedly the most powerful and profound of the treaty's stipulations:

> **ARTICLE V**
>
> 1. Any nuclear explosions in Antarctica and the disposal there of radioactive waste material shall be prohibited.
>
> 2. In the event of the conclusion of international agreements concerning the use of nuclear energy, including nuclear explosions and the disposal of radioactive waste material, to which all of the Contracting Parties whose representatives are entitled to participate in the meetings provided for under Article IX are parties, the rules established under such agreements shall apply in Antarctica.[21]

This means that for the past sixty-plus years, any spot on planet Earth south of the latitude 60° south parallel has been a nuclear-free zone. What's more, there are now fifty-seven nations signatory to the Antarctic Treaty, all committed to the peaceful protection, preservation, and study of the seventh continent. That is perhaps the greatest triumph of IGY.

*   *   *

More about Antarctica: IGY's activities in Antarctica got a lot of press, but I'll close this chapter with a brief summary of the objectives and results from some of the lesser-known, smaller surprises of the IGY Antarctica program, taken from the 1965 *Report on the U.S. Program for the International Geophysical Year*:

Summary of Antarctica Activity Program

Six major scientific stations were established in Antarctica to carry out the U.S. scientific program during the IGY. Private institutions and government agencies (concerned with scientific research) conducted scientific research at these stations. Logistic support for the Antarctic program was furnished by the Department of Defense.

Cosmic Rays

Objectives: The program augmented the U.S. and global cosmic-ray program for exploration of the variations in mass and energy of the primary cosmic radiation and the recording of variations in cosmic-ray intensity.

Preliminary Results: Data have shown that the seasonal variation of the secondary cosmic-ray activity, as measured at ground level, is probably due to atmospheric conditions rather than to variations in the primary cosmic-ray flux.

Geomagnetism

Objectives: Studies contributed to a fuller understanding of the earth's zonal current systems and determination of a more reliable location of the southern auroral zone. In connection with other studies, particularly aurora and ionospheric physics, geomagnetic studies permitted investigations of the time relationship and spatial comparisons of simultaneous disturbances in both the Southern and Northern Hemispheres.

Preliminary Results: These data have emphasized the inter-disciplinary nature of upper atmosphere disturbances. In addition, data in both the arctic and antarctic have shown evidence of a disturbance called the "cuped bay" which should be investigated as a supplement to the use of hourly values.

Glaciology

Objectives: The main objectives of the U.S. glaciological program in antarctica were to gather information on the volume of antarctic ice, the topography of the ice surface, and the land beneath the ice; ascertain the present status of the antarctic ice sheet (whether it is gaining or losing in mass and volume and the manner in which this gain or loss is taking place); obtain information relating to the history of the antarctic ice, determine its trend and how it will react to changes in solar radiation and ocean temperatures. To accomplish these objectives an extensive glaciology program was carried out including station programs, deep core drilling, ice deformation studies and major oversnow and airborne traverses.

Results: Studies reveal that previous estimates of the volume of the antarctic ice cap have been underestimated, possibly by as much as 40 percent. Temperatures taken by the oversnow traverses at ten-meter depths have been demonstrated to be closely equal to those of the average annual surface temperature. Results concerning the sub-ice topography of Antarctica show a great channel between the Ross & Bellingshausen Seas, better definition of the boundaries of the Ross and Filchner Ice Shelves, and discovery of an extensive area in Victoria Land, West of the Antarctic Horst, where the rock surface is below or close to sea level.

Gravity

Objectives: An extensive gravity program was carried out during the IGY in Antarctica to form a common and permanent gravity base for use as a reference for the gravimetry programs of the United States and those of other nations. In addition, gravity measurements were used to augment the seismology ice thickness programs.

Preliminary Results: Observations show that the polar ionosphere retains considerable ionization throughout the dark period of winter and, moreover, shows a weak diurnal variation even though the sun remains at a fixed angle below the horizon. Whistler and VLF recordings were observed [at] Byrd Station more frequently than at corresponding northern magnetic latitudes, presumably as a consequence of the very low noise level in Antarctica. In addition, it is inferred that the frequency characteristics of the signals are dependent upon magnetic latitudes.

Interview with Louise T. Huffman, MSEd

Director, Education and Outreach for the U.S. Ice Drilling Program Office, Dartmouth College; Chair, Formal Education Subcommittee of the 2007–2008 International Polar Office (IPY) Education and Outreach Committee for the International Polar Year

Mark O'Connell: If scientists around the world could drop their differences in 1957 and all work together to solve some of the biggest mysteries of the earth, could they still do that now to tackle climate change? Is that naive? Or is that actually something that could happen?

Louise Huffman: I honestly think scientists are working together around the world on climate change. But there's been a real campaign of misinformation. And so there're a lot of people that don't believe in climate change in the United States because of that. And it's funded by fossil fuel industry. And they've done I think, some, hopefully not, irreparable harm, but there are too many people that don't understand that this is real, and it's happening now. And it goes all the way up through our politicians and government. So the scientists do work together. I mean, there are lots of international studies going on about the climate. In Antarctica, you have to cooperate and you have to share everything. That's the Antarctic Treaty. So if you're working in Antarctica, you don't have an option on that. You share data, you share resources. If your countries are at war back home, you're not when you're in Antarctica, so I don't think it's the scientists.

M.O.: How do you push back against that misinformation?

L.H.: We're doing a conference for middle school and high school teachers in the Chicago area called Climate of HOPE, and it's going to be pretty amazing. We have some great speakers, keynote speakers. Frank Leopold works for NOAA, but also has been an author on the new *National Climate Assessment Report*, which is an amazing document.

And it's granular, it digs down into regional areas, and what does the science say is going to happen? We're working with tribal colleges. We have Amanda Townley coming in; she's the executive director of the National Center for Science Education, who spent a lot of time working to let teachers know what is misinformation. Example, the Heritage Foundation sent out tens of thousands of these really slick-looking teachers' guides for teaching climate change, except it was all funded by the fossil fuel industry, and all wrong and not based on science. And so the National Center for Science Education spent a lot of time and effort to let teachers know what was incorrect about it to make sure they were aware [of] these things and this wasn't something they should be teaching.

M.O.: How did you become involved with climate science and IGY? Was it always of interest to you?

L.H.: I can't pretend to be an expert on the IGY, although I did play a role in the 2007–2009 International Polar Year and have taught about the IGY when I was a teacher. As a young student during the IGY, I actually remember hearing about it and believe some of its fascination may have led me to becoming deeply involved in polar and climate science education. I did read about it in the *Weekly Reader*!

My fifth-grade teacher Mrs. Morgan talked about it, and she had obviously been affected by it and brought it to us. She had some printed materials that had come from IGY that she had kept. She is the teacher, I think, that probably turned me on to science, the most of any teacher. The fact that they had put money into education through IGY to teachers, I think was probably a really novel idea at the time. And IGY was earth shattering, shaking. When I went to college, they really were beginning to teach about plate tectonics. But they had learned about it during IGY; they had found the evidence that yes, this is what's happening. So that theory of plate tectonics developed out of IGY, and the Van Allen radiation belts. So much of outer space exploration came because of IGY; all of that affected me and it was probably the first time that I realized that science is something that you do. Science was happening right now. And we didn't know the answers. That, I think, opened my mind to the idea that science at that point was something that you do.

M.O.: You had mentioned earlier how unique and wonderful it was to have scientists from many different countries all working together during IPY in this sort of bubble of Antarctic science. Is that a model for other sciences? Do you see that happening in other sciences, or is this totally unique to the glaciologists?

L.H.: So there are all kinds of different sciences within ice core science. Because during IPY, for example, there were all kinds of scientists, even social scientists, that were studying different aspects of the Arctic and the Antarctic. So there was a lot of collaboration and cooperation going on. And I think the scientific world works differently than the political world, for sure. I think you get more and better questions from a more diverse group that's asking the questions and looking for the answers. I know geoscience tends to be not very diverse. And we're working hard to try to change that. A lot more women are in geosciences now than there were, so that's one step forward. But you know, it's the old picture of the scientist and the bearded white man and a lab coat, and we're trying to really get away from that, getting more women into geosciences, getting more people from different cultural backgrounds into geosciences. And that's one of the reasons that we're hoping to work with the tribal colleges too, because that indigenous knowledge is a whole 'nother piece that's really important, and how do you look at climate change through two sets of eyes, the Western science and the indigenous knowledge science, so it's pretty fascinating. Yeah, so diversity is really important, and if you're just working with people that think like you, you're not going to get the best ideas. You'll get more and better ideas when you put people from lots of different backgrounds together to think on the problem.

CHAPTER 5

The Auroras

4 Large Incoming Solar Bursts Could Supercharge the Auroras This
Weekend

We Have a Very Rare Event on Our Hands

Auroras Everywhere!

Historic Geomagnetic Storm Dazzles

Auroras on May 10–11 Wowed Millions!

Jaw-Dropping Northern Lights from Massive Solar Flares Amaze Sky-
Watchers around the World

We May Have Just Witnessed Some of the Strongest Auroras in 500
Years

These headlines from across the internet that appeared before, during,
and after the great northern lights spectacular of May 10–11, 2024,
tell the story. Something big was happening with our atmosphere and
our sun that night that was destined to give millions of people a rare
and wondrous look at the northern lights. People as far south as where
I live in Atlanta were promised an amazing light show, and countless
sky watchers were able to enjoy a transfixing electric light display wav-
ing and rippling overhead in the wee hours.

My wife and I, alas, had no luck. We went out in our car in search
of the aurora but couldn't escape our city lights. We were also denied a
show at home, as our house is tucked away in a small clearing in a big

forest and does not allow for much in the way of stargazing (I rarely, if ever, break out my telescope, for that very reason). So, a disappointing night for us, especially when we heard reports the next day that there had been multiple auroras visible in places not all that far from us. We had given up too quickly.

Just the same, I felt some measure of excitement regarding the aurora's southernly intrusion, knowing as I did that I would soon be writing a chapter on this very topic. Because who isn't fascinated by these ghostly, gauzy veils of color and light floating silently in the night sky? Also, I had already seen the northern lights—in the north, of all places. Wisconsin, to be exact. Twice. So I already knew what it felt like to behold them. The first sighting occurred on a late-night drive to visit family, and the second sighting occurred several years later while I was camping with friends on a small island in Lake Superior, in what we Wisconsinites colloquially refer to as "Up Nort."

What I saw on both of those occasions were sheets of fuzzily glowing, slowly rippling incandescence high in the firmament that shone a soft mixture of white and red and stretched out across a good 25 or 30 percent of the night sky. The lights seemed to want to touch down on the ground, like some kind of slow, cool lightning, but something held them up, dangling them above the earth, never letting them reach the ground. The first time I was able to enjoy the sighting while pulled over along a country highway with my young family. (We tried to wake the kids, but they were zonked.) The second encounter, the one that took place while I was camping on the island with friends, lasted longer: Two of my camping buddies and I were having our last beers of the evening as the campfire died down, and we watched in wonderment for an hour or so, talking in hushed tones as the lights glowed on.*

---

* These days, if you're an adventurous traveler, you can trek north of the Arctic Circle and spend a few days and nights in the land of the midnight sun basking in the glow of the northern lights.

Of course, you're taking your chances. Just as there's no guarantee on an Antarctic cruise that you'll actually set foot on the continent, what with the unpredictability of both the sea and the weather, there is no guarantee that the aurora will appear during your sojourn to the north. But the odds are in your favor if you plan your travel well: the auroras—both northern and southern—are "in season" between late August and mid-April.

*   *   *

People have been talking about the lights in hushed tones since prehistory, but it wasn't until the early seventeenth century that the lights were subject to serious scientific study. This is when Galileo, one of the most influential and controversial scientists of his day, turned his attention to the phenomenon and began to unravel the mysterious forces that create them. It was Galileo who gave the lights their formal name, the aurora borealis, named after Aurora, the Roman goddess of dawn, and Boreas, the Greek name for the north wind. From his studies Galileo surmised that the aurora was caused by sunlight reflecting off the earth's atmosphere, and that explanation was accepted until the first International Polar Year of 1882–1883. During this inaugural IPY, a number of observation posts were established around the globe, allowing for the first long-term, comprehensive record of auroral appearances and behavior.

"Distant as it is, the aurora is important to the world-wide effort of the IGY," Alexander Marshack wrote in 1958, "for it tells more dramatically than any other phenomena of the relationship between the sun and the atmosphere, and between the sun and the Earth's magnetic field."[1]

By the time of IGY, scientists were able to incorporate a new discovery into their theorizing, Marshack revealed. "At the time of IPY, scientists studied only the northern aurora," he wrote. "IGY scientists are studying both the Arctic and Antarctic auroras."[2] So not only did the population of earthly auroras double between 1883 and 1957, but those northern and southern auroras were also found by IGY scientists to appear simultaneously! At the exact moment on May 10, 2024, when millions of people in the Northern Hemisphere were witnessing

---

One aurora resort situated on a small, frigid island in the northern reaches of Norway recommends visitors come for four to seven days for their best chance to see the lights. "Ps: It is impossible to say what is the best time or month to see the northern lights," the website cautions, "as this depends on the weather." After pointing out that any time from September to April can be equally opportune for aurora viewing, the proprietors inform the potential sky watcher that the viewing is equally good whether the moon is full or new. "This is subjective and we personally like both equally well," they say with great diplomacy.

the appearance of the northern lights, millions more were witnessing the appearance of the southern lights south of the equator. Uncanny.

"Among all earthly phenomena one of the most eerie and spectacular is the aurora Polaris," wrote J. Tuzo Wilson,[3] who estimated that the aurora might appear once or twice over Europe and the United States every ten years and might be visible as far south as the tropics once every century. The southern auroras, Wilson adds, rarely venture as far north as Tasmania and New Zealand.

*   *   *

It's not a surprise, then, that the auroras are integral to the mythologies of many cultures around the world. "Human habitation of the North dates back tens of thousands of years, and with Artic occupation come stories of the aurora," wrote Melanie Windridge in *Aurora: In Search of the Northern Lights*. "The delicate patterns and striking colors grab people's imagination like no other natural phenomenon. The lights have featured in art, poetry and literature throughout the centuries, and they often have an almost spiritual effect on those they touch. As part of the fabric of northern lives, the aurora borealis features in folk tales and legends, sometimes in a starring role, sometimes as a by-the-by."[4]

Because the aurora often glows red, for example, it has at times been associated with both fire and blood and thus has been seen as an evil omen. One tale tells of a Roman legionnaire who saw a red aurora and feared that his city was burning down. Others tell of red auroras presaging both the start of the U.S. Civil War and the assassination of Julius Caesar.

Alexander Marshack recalled how the sight of a blood-red aurora caused a Roman historian named Livy to report that "the heavens were seen to blaze with hidden fires. . . . [T]o avert the alarms of the terror-stricken, a three-day season of prayers was ordered."[5]

Norse mythology often portrays the aurora as spirits of the dead—sometimes evil, sometimes benevolent. Some are said to be spirits of dead relatives, others of fallen warriors. Some stories claim that the

lights represent animal spirits: swans, foxes, and even herrings.* "Old
Norse mythology speaks of the aurora as reflections of light from the
shields of the Valkyries, the beautiful virgin warriors who rode over the
battlefields choosing the slain," Windridge wrote.[6]

In his 1960 book, Sydney Chapman reported of the northern
lights: "In earlier times, many fanciful interpretations were put upon
them. They were sometimes pictured as fiery horsemen contending
in the sky. In parts of the north of England they were known as Lord
Derwentwater's lights, because long ago a brilliant display coincided
with—and was counted as an omen of—the execution (for armed
rebellion) of a nobleman of that name."[7]

Far more eerie was a particularly spooky manifestation that occurred around huge swaths of the globe in September 1859. Scientists
named it "The Carrington Event" after the amateur British astronomer who witnessed the CME that triggered it. The event began as
Richard Carrington was drawing an image of a growing number of
sunspots, as was his hobby. A brilliant flash seeming to originate from
the sunspot he was sketching caught his attention, and he made note
of it, observing that the "flash" had a duration of several minutes. The
very next day, the earth was struck by the largest magnetic storm in recorded history, twice the size of the previous record holder. Carrington
put the pieces together and surmised that the flash and the storm were
linked and that in fact the former actually triggered the latter.

The spectacular auroral display during the storm was seen and
experienced by millions who had never before imagined such a sight,
and those millions had quite a story to tell. The magnetic storm was so

---

* Of special interest to me is the Norse myth in which the aurora was believed to be the
pulsating "Bifrost Bridge," a radiant arch that led fallen warriors to their final place of rest
in Valhalla. Why does this myth appeal to me the way it does? The Bifrost was a central element in an episode of the early 1960s science fiction anthology series *The Outer Limits*. In
the episode "The Bellero Shield," an alien from another world is brought to Earth and into
contact with humans by way of a glowing, pulsing beam of light that one character describes
in reverent tones as a Bifrost. Little did the younger me watching that episode realize that the
mysterious Bifrost that transported a benevolent alien to Earth was in actuality the northern
lights! The screenwriter also tapped into Greek mythology with the title of the episode. Bellero
is short for Bellerophon, a Greek "slayer of monsters" who defeated the Chimera and captured
and trained Pegasus, the famous winged horse.

intense that scientists did not have an accurate measurement of its size, because there were no measuring instruments in existence that could register readings that high. They were literally off the charts.

"All over the planet, as far away from the poles as South Carolina and Cuba, bright colourful auroras illuminated the nighttime skies," reported the South African National Space Agency. "It seemed that the sky was on fire."[8] Indeed, some early risers thought the sun had risen without them. As it drew closer and closer to the lower latitudes, that "fire" found an attractive means of travel, as charged particles from the sun bonded themselves to telegraph wires, where they caused no end of havoc. These wires, stretching in the United States along the Eastern Seaboard, became supercharged, emitting showers of sparks, burning up transformers, charring papers, injuring telegraph operators, and—here's the spooky part—inexplicably coming to life and sending telegrams long after the electric power had been shut off.

"When the earth's magnetic field fluctuates during a geomagnetic storm, the process induces a current along distances of long conductive materials," wrote Dr. French in his book about the sun. "Unfortunately for us, our power lines, telegraph wires and railroad networks, as is their design, are built of exactly these."[9]

As a result of the storm, large segments of the country went dark and silent, as bewildered telegraph operators found they could no longer send or receive telegrams because their batteries were rendered useless by the auroral currents. But in a bit of *Ripley's Believe It or Not!* weirdness, when the valiant operators disconnected those dead batteries, their telegraphs suddenly came to life with no apparent power source. One telegraph operator in Boston informed his counterpart in Portland, Maine, that the only way he could get any business done was to disconnect the batteries and let the auroral current shuttle the messages along the telegraph wires. When asked by the Boston operator how he was getting along without power, the Portland operator replied, "Better than with our batteries on . . . Shall I go ahead with business?"

"People began to realize that there may be a darker side to the northern lights," wrote physicist Melanie Windridge about the emerging science of "space weather."[10]

*   *   *

If a solar storm the likes of the 1859 Carrington storm descended on us today, interference from the auroral currents could damage, even disable, many of the technologies we have learned to take for granted. Power grids, communication and weather satellites, cell phones, nav systems, and, yes, even telegraph wires could all be disabled and cause untold social and economic disruption. At least today science has developed measuring technology that will be able to determine the full power of another Carrington Event, should one occur. And one could.

"Distant as it is, the aurora is important to the world-wide effort of the IGY for it tells more dramatically than any other phenomenon of the relation between the sun and the atmosphere, and between the sun and the earth's magnetic field," wrote Marshack in 1958. "Since auroras are most common at the peak of a sunspot cycle, the geophysicists in 1957–58 have a rare opportunity for studying this high atmosphere as well as the relation between the sun and the auroras, and between auroras and magnetic ionospheric disturbances."

"With the expansion of the Canadian frontier northward, the aurora has created new problems in communications," he cautioned. "For example, the auroral zone coincides with the North American Arctic defense barrier, the DEW (Distant Early Warning) Line, completed in 1957 and designed to alert us to Soviet aircraft or missiles approaching from over the arctic and then across Canada. Additionally, the aurora affects radio, radar and the guided missile program along this defense perimeter running from Greenland to Alaska."[11]

Here's a simple way to understand what causes the auroras, courtesy of the website of the Royal Museums Greenwich, part of the UK's Royal Observatory in London:

The lights we see in the night sky are in actual fact caused by activity on the surface of the Sun. Solar storms on our star's surface give out huge clouds of electrically charged particles. These particles can travel millions of miles, and some may eventually collide with the Earth.

Most of these particles are deflected away, but some become captured in the Earth's magnetic field, accelerating down towards the north and south poles into the atmosphere. This is why aurora activity is concentrated at the magnetic poles. What we are seeing therefore are atoms and molecules in our atmosphere colliding with particles from the Sun. The aurora's characteristic wavy patterns and "curtains" of light are caused by the lines of force in the Earth's magnetic field.[12]

Prior to IGY, the lights were a vast mystery, a startling, often haunting manifestation of nature that was visible to a privileged and hardy segment of the human race and that inspired both awe and fear in their hearts. To place the auroras under the microscope, so to speak, it would be necessary to keep not just the poles but also the entire planet under constant observation for the duration of IGY, no small feat. This program of surveillance would not have been possible during the relatively primitive First and Second Polar Years; for example, photographs of the aurora taken during the first IPY were deemed unsatisfactory, as they were calibrated on the assumption that the aurora formed at an altitude of five miles, when it was discovered later that it actually formed at an altitude of sixty miles.

But by the time of IGY, imaging technology had progressed dramatically. Aurora scientists at the Geophysical Institute of the University of Alaska, for example, developed a special domed camera programmed to take full-sky pictures on a regular schedule: some every minute and others every five minutes. These photographs would go a long way toward determining the degree to which the two auroras resembled each other in scale and hue and appeared at opposite ends of the earth at the exact same time.

"In high latitudes, both Arctic and Antarctic, a network of about 120 stations was organized, for the photographic record of the aurora," Chapman wrote in his 1960 book. "These cameras produced an almost embarrassingly abundant photographic record of the aurora, that will take much time to study adequately."[13]

There was another overwhelming source of auroral data in operation during IGY, though. "A great effort was made to recruit and organize and train visual observers to watch for and record the auroras that might come within their view," Chapman wrote.[14] Over the course of IGY, groups of eager volunteer sky watchers formed in the United States and Europe, as well as in Mexico, Central and South America, Australia, New Zealand, Africa, and Asia.

"Auroral observations probably involved more amateur enthusiasts than any other IGY effort," wrote *New York Times* reporter Walter Sullivan in his 1961 book. "They provided a picture of the displays more extensive than any previously possible. It is easy enough for one man to look up and see an aurora, but to combine the sightings into a global pattern was the unique contribution of the IGY."[15]

"The Auroral Data Center at Cornell received about 18,000 hourly reports a month from some 430 American sky watchers, including 30 airline pilots and an assortment of seafarers, amateur astronomers, science teachers and the like," Sullivan added. "In addition, 130 stations of the United States Weather bureau made observations every hour. The report cards came in almost equal numbers from the volunteers and from the weathermen."[16] Wilson observed that the IGY program would produce "literally several million" reports and "a few million feet of photographic film."[17] Safe to say that never before had so many earthlings simultaneously observed a display of the aurora.

*   *   *

Leading up to the start of IGY, Marshack anticipated another, more spectacular source of aurora data: "Two hundred U.S. rockets," he wrote, "are being sent up from all parts of the world, to study the upper atmosphere at heights of nearly 200 miles. Of these 200 U.S. rockets, 115 will be sent up to conduct experiments on and measurements of the auroral zone, that is, more for the aurora than for any other phenomenon of the upper atmosphere."[18]

It's curious, then, that these vast amounts of data collected by volunteer skywatchers required the use of an amazing new scientific device, the "counting computer." "One of the mid-century developments that offered hope for digesting the mountains of data assembled in the IGY was the advent of business machines and computers. Nowhere were these found more essential than in the aurora program," Sullivan wrote. "In preparation for the IGY, 'mark-sense' cards were designed on which information could be entered by darkening certain squares with a pencil. These cards could be processed entirely by machine, the first step being 'automatic reading' of the pencil marks and punching of the information in standard business machine fashion."[19]

In an early test of the computerized recording system, Carl Gartlein, the director of the data center managing the effort, used "machine analysis" to determine whether the aurora traveled from the east to the west in lockstep with the sun, and whether the average motion of the aurora reversed at midnight, as was supposed. "The results surprised us," wrote Gartlein in his newsletter. "1) The predominant average motion of auroras is toward the east, in the direction opposite to that of the sun and the stars. 2) Although the average motion of an aurora depends on the time, there is no reversal at midnight."[20]

"There are many visible forms of the aurora—rays, flames, glows, arcs, bands, 'draperies' and coronas. Some auroral lights are diffuse and shapeless; others are as steady as a searchlight beam," Marshack noted. "An aurora may flash across the sky for a second, or it may last a full night in a grand, ever-changing display." "The colors are subtle," he went on. "They may be pure white, blood-red, yellow-green, blue green, blue, violet or even gray. They are caused by the excitation of nitrogen and oxygen in the thick ionosphere."[21]

"The aurora is in many respects still as mysterious as it is beautiful," Chapman wrote in 1960. "By recording its form, color and brightness from many places, it is possible to determine the true location of its luminosity in the atmosphere—its height and its geographical position in plan."[22] "The results," Marshack reported, "may reveal facts about the earth, outer space and the solar system that are not yet suspected."[23]

This is one good reason to study the auroras, both of them—borealis and australis. As it turned out, the auroras were a priority for the planners of IGY. It's clear from the planning report that when the "Aurora and Airglow" panel began planning its IGY research activities, they left no proton unturned, as reflected in the following list:

a. *Visual observations.* The panel recognized that visual observations would be an important element of U.S. participation, inasmuch as the auroral zone dips sharply southward over northeastern North America. The panel planned to collaborate with Canadian auroral scientists in working out a scheme of recording visual observations. Different standards of observation were set according to whether the location was southerly (where auroras would not occur frequently) or the station would be northerly (where auroras would be very frequent).

b. *Photographic observations.* The panel planned to utilize a camera that would photograph the entire overhead hemisphere so that wide synoptic coverage could be obtained of auroral incidents and development. They anticipated at that time that photographs would be taken at one-minute intervals and read at fifteen-minute intervals for synoptic observations. They also thought that all except a few key stations would observe only on World Days.

c. *Radar observations.* In spite of problems of interpreting radar echoes and distinguishing auroral echoes from echoes from other ionospheric phenomena, the panel decided to provide a longitudinal chain of stations at approximate geomagnetic latitude 60 degrees along with two trans-auroral networks, one in Alaska and the other in the mid–United States connecting with stations in Canada.

d. *Conjugate point observations.* The panel thought it was possible to institute coordinated observations at the ends of magnetic field lines in cooperation with scientists of Australia and New Zealand. These would include visual, photographic, photometric, radar, and spectroscopic observations.

e. *Spectroscopic observations.* In order to provide information on the distribution of incoming solar particles thought to be responsible for the auroras, the panel planned to set up a network of stations using a Memel patrol spectrograph. This instrument would focus a 165° arc of the sky on the slit of the spectrograph. The camera, oriented in the magnetic meridian, could thus provide a distribution along that meridian of radiation, and from that the distribution of the bombarding protons could be deduced.

f. *Spectrophotometric studies.* The same spectrographs mentioned previously could be used with proper calibration to observe ratios of intensities of certain line and band systems, which are related to the physical processes involved in auroral excitation.

g. *Scintillation of radio stars.* The panel thought that observations of scintillation of radio stars, using a tripartite observing network, would provide information on the size and velocity of patches of ionization that might be connected with the auroral display.

h. *Absorption of extraterrestrial radio signals.* Initial experiments at the Geophysical Institute in Alaska indicated the usefulness of observing the variations in the galactic noise received at the ground. The variations come about as a result of changing absorption in the ionosphere and are related to polar blackouts, geomagnetic and ionospheric storms, and auroral activity. The panel thought that a frequency of about 30 megacycles would be useful.

i. *Radio noise associated with auroral displays.* There had been some indication of 3,000 megacycle radio noise emitted by auroral displays. This general question was under study at the Geophysical Institute in Alaska, and the panel expected that if the results were promising, several stations should be set up to monitor this phenomenon.

j. *Airglow studies.* The panel expected to institute a chain of synoptic stations to monitor the several emissions of the airglow with photo-electric photometers. A meridional chain would be set up through North America connecting to stations in South America.

A full quarter century after the end of IGY, scientists still couldn't explain how or why the auroras appeared. "Fundamental questions are still unanswered," said physicist Herbert Friedman in a 1983 oral history. "We don't know what makes the aurora. We know of all sorts of possibilities. Can somebody devise a critically definitive experiment which will tell you that the aurora is generated by a large dynamo process of solar wind flowing over the polar cusps, or is it the result of some tearing-mode instability in the geomagnetic tail? Those are two strong possibilities. Somebody's got to find the answers."[24]

*   *   *

More about the auroras: Although it was big news that the northern and southern auroras were somehow in sync with each other, I end this chapter with a brief summary of the objectives and results from some of the lesser-known, smaller accomplishments of the IGY aurora program, taken from the 1965 *Report on the U.S. Program for the International Geophysical Year*:

Summary of Auroral Activity Program
The aurora program consisted of observational networks based on the all-sky camera, visual observations, spectrographs, radar and cosmic noise recorders. Some projects were established for area operations and reduction of data.

All-Sky Camera Construction
Objectives: This project provided for the development and construction of a camera to photograph the complete overhead hemisphere (hence the term "all-sky camera") for use at U.S. aurora stations and at certain stations operated in cooperation with other countries.

Operations: A prototype design was made in 1955; of several cameras built then, three continued to operate successfully in Alaska during the IGY, supplementing the more elaborate camera constructed under this project. For the final design, the prototype was modified to conform with an instrument constructed by the Swedish scientist W. Stoffregen, in which a folded optical system resulted in a more compact design. Twenty-nine cameras of this improved design were constructed under contract by Photomechanlsms, Inc.

Visual Aurora Observations

Objectives: This project at Cornell University provided for systematic synoptic visual observations of the aurora in the United States, Alaska and the Antarctic, and assured the coordination of observations both in the United States and Canada.

Results: Newsletters were sent to all observers and about 50 other interested persons from the beginning of the IGY to date. The letters contain additional hints on observing, results of observations and preliminary analyses. Preliminary analysis shows a close relation between the magnetic K-lndex and location of the overhead zone of aurora. It also shows a diurnal effect causing auroras to be farthest south at midnight.

Patrol Spectrograph—Prototype Development

Objectives: This project at Yerkes Observatory, University of Chicago, was initiated to provide for the development of a patrol spectrograph with automatic operation that could then be manufactured under contract for the U.S. station network.

Operations: The design of the spectrograph was carried out at Yerkes and a prototype was developed. Specifications were prepared for the manufacture of a number of instruments and are as follows:

- Collimator, Petzval, Two Separated Doublets
  - Focal Length, 24 inch
  - Working Aperture, f/8

- Transmission Grating, 600 lines/mm, blazed for
  - 1st order at 5100 A
- Camera, Semi-Solid Schmidt
  - Aperture, 3-inch, f/0.625
  - Dispersion, 1st order, 350 A/mm

Testing of the prototype was done at Yerkes. Observers for the Antarctic were trained at Yerkes in the use of the spectrograph.

Aurora Data Acquisition and Processing

Objectives: This project at Cornell University was set up to reduce to punch card form data from all-sky cameras and visual observers, prepare synoptic maps and statistical tables, and distribute large numbers of multilith copies.

Operations: Visual observations were sent to Cornell in diagram form and also in mark-sense (computer readable) form. Where data were not supplied in mark-sense form, punch cards were made and processed. Lists of data were run off the punch cards and submitted to WDC-A: Aurora (Visual) for transmission to other WDCs.

Interview with Meredith Goins, MSIS

Executive Director, World Data System (WDS)
International Program Office (IPO), University of Tennessee–
Oak Ridge Innovation Institute

Mark O'Connell: I find it amazing that after nearly seventy years the data from IGY is still preserved by the World Data System. Can you tell me how all that data is protected and utilized today?

Meredith Goins: I'm the executive director of the World Data System International Program Office. WDS consists of two offices, both the program office that I run, we're based out of the University of Tennessee, but we're located physically in Oak Ridge National Labs. Our partner, our International Technology Office, is based out of University of Victoria in Canada. And so I run the organization operations, and the Technology Office runs more of the research and hands-on support to the repository.

M.O.: Is there a documented lineage between the World Data Centers (WDCs) of IGY and the World Data System you manage today?

M.G.: I'm glad you asked, because one of the things I have my admin doing right now is we're reviewing and have been for the last year all of our Memos of Understanding (MOUs). So, our former MOUs were with ICSU, which is the precursor to the International Science Council (ISC). And because we were also housed in Japan and we've moved to the United States, we've asked everybody to renew their MOUs because the organizations have changed the governing body of the ISC, which oversees our existence. We're an affiliated body, they're the ones that allowed us to be available when the IPO office went up for renewal after ten years of very effective and excellent service in Japan. So, just to explain the structure, the WDCs fell in line with us, and many of our centers are now housed under one entity. I'll give you an example: the National Oceanographic and Atmospheric Administration (NOAA) in the United States hosts three or four of them.* I know China hosts a few of the WDCs.

All members have to agree with our data policy principles and also become a core trust seal certified repository. I say this because that's a peer review of the repository. And we know how important peer review is in science, having another like entity review to make sure that you're maintaining your data. Well, basically, you're doing what you say you're going to do.

So that's one of the reasons why we got a certification there. And we did that to make sure our WDCs were maintaining their data well, and building those partnerships. The sustainability of data requires us to have partners.

If I hold a data set, I should also have somebody else who has a backup. So let's say I need to shut down the WDC—an earthquake shuts me down, a tsunami, anything, you know, it could be financial, it could be political, it could be hazard from Mother Nature. So having that backup plan, that's where the WDCs are so important, because they support each other: "I need you to host this data, while I rebuild after the earthquake," that kind of thing. So, there are those plans of how they back up things. In the old days, we used to take reel to reel tapes and ship them to different places. Now we just put it up to the cloud and give access to another entity for them to be able to maintain while we do whatever needs to happen in whatever crisis has occurred, if that makes sense.

M.O.: Has there ever been such a crisis?

M.G.: There has, and the International Science Council actually detailed this in a document recently. It's from their Centre for Science Futures. In this document, it talks about data in times of crisis. And there're a lot of different examples of this: Ukraine, we know what's happening there, and that actually an entity in the UK was their backup. And so they were able to support that data, because that data center no longer even exists physically. But the data still lives because there was a backup plan. Right?

---

* NOAA hosts the World Data Services for Geophysics, Meteorology, Oceanography, and Paleoclimatology.

M.O.: Who uses the data that your organization is safeguarding? Or maybe who *has* used it, maybe a little historical perspective on that as well.

M.G.: Who typically uses it, I can answer. Governments use this data. Scientists use this data. Publications use it. So I would have to reach out to every single one of the centers to ask them for that. And how many centers are we talking about? Well, that depends. Yes, so if you looked at the initial data sets, initial centers, some of them no longer exist, and I can't determine what has happened to them. And then there's others that still exist and are out there, but do not respond. Sometimes language is a barrier.

M.O.: What involvement do you have with ISC? Do you work with them on a regular basis?

M.G.: So our action plan on our website is directly linked to what ISC is doing. We look at objectives and action supporting their strategies and align with their action plan as well. And those are digital transformations of science and education, which makes sense. And then digital technologies for sustainability. And sustainability doesn't just mean making sure that we're using our resources. It also means the sustainability of the data sets, and the knowledge: How do you document and maintain those knowledge pieces?

M.O.: Maintaining knowledge, how do you do that? Is there a simple way to explain how you go about doing that?

M.G.: By encouraging our members to hold their data in long-term data repositories, that gives the ability to show proof of what has happened before. The data sets, being able to see how long something has been happening or not happening, helps build that knowledge over time. If you can see that big picture, you can look back and say, you know, there has been change, there has been climate change, or there hasn't been climate change; it depends on what you're looking at, and how you define climate change. There's a lot of ifs in there. Depends on what your process is and what your interest is in. Welcome to Science, right?

So we care very deeply, scientists care very deeply about reusability, interoperability. So if I'm going in, and I want somebody to validate my findings, they need my data. And a lot of papers say, please reach out to the corresponding author to get the data—and guess how many responses you'll get from that? Not very many. But everybody wants to see your data, actually. So by having it already linked into a data repository, with a static Digital Object Identifier, a DOI that then allows people to find that data, it allows people to then play with it and see if they have the same reproducible findings.

And that's quality science. That's how we judge science. That's how we maintain knowledge. And by having access to that data to— and we're not doubting each other; we're trying to often reduce science so that we can add to it—it's that building block, and the data is a critical component of that building block.

M.O.: So about the International Science Council, I was intrigued by their mission, the Mission for Sustainability. They talk a lot about establishing data hubs. And when I started reading about those, I immediately thought, well, that sounds a little like a World Data Center from IGY. Am I off track there, or is there some similarity?

M.G.: There is some similarity, but they're also looking at different things as well. For example, we only look at scientific data. Now scientific data does incorporate the humanities, it incorporates earth sciences, it incorporates economics, and it does more than just our sciences, or physics, for example; it is all fields. And we've recognized we need to do bigger and better and include more, such as the digital humanities, in this, because the context behind that is that the humanities often give that picture, a more vibrant picture, of the research. So not just here's what's happened, but then it's look at here's how it is used in life. Here's a snapshot of how others have created art because of this. There's a lot of connections there.

I was talking about the Center for Science Futures. [Their paper] talks about Ukraine, things that happen in Sudan, and Gaza, the destruction of a library in Cape Town. All that information is gone if there's not another copy somewhere. So we're trying to be preventative and make sure that this information is available. So we can continue. We don't want another Library of Alexandria happening. We'll never know, because it's lost, right.

And the other thing, from an economic standpoint, doing this really helps show funders—for example, the U.S. government—they allocate billions of dollars every year on research, and they can now see an outcome and can track how it's built upon. So it's a measurement, if you will, but also it's showing that our government is being effective in releasing dollars and making things happen, instead of just "Oh, yeah, go do this science, have fun." Not that that's what they did. But you know what I'm saying. So it helps with that aspect as well.

M.O.: With the fifth International Polar Year coming up—IGY being the fourth—I'm wondering if you have any thoughts on why these scientific events are still of value to us today?

M.G.: I'm going to give you a personal opinion, not a WDS opinion. International Polar Years are phenomenal. You have a group of people from all over the world coming together. And that's very hard to do in science, to get everybody to agree on what we're going to capture, when we're going to capture it, how we're going to capture it, and then where we're going to store it. Having that level of agreement is hard. It's very hard.

And so being able to do that repetitively is huge, because the field, which is actually a whole lot of fields, have come together and said, this is important. So they've started announcing that so that people can start planning.

People are already trying to figure out what and how do we want to be involved? What do we want to do? What's our drive to participate? And how do we get our local funders to agree that this is a worthwhile cause? This is really expensive. So you're going to have to hire people, you're going to need resources, not just money but boats, you're going to—excuse me, ships, they're not boats—somebody in the navy would be very mad at me right now.

Thinking of all those things, the different technologies you need to do the measurements, the storage space, all of the documentation that's required, which involves hiring staff and possibly physical locations, lodging, all of that. What does that look like? That's a lot of funds throughout the world. So people have to agree and sign on early to plan for such an effort. And we're building unity. Some organizations call it science for peace or something like that. But it's us coming together. And the reason I really agree with the ISC's mission of science for the greater global good, this very much hits that as well. And that's what our data repositories do is help make sure that that science is available for everybody. Remember that not all countries have equal access to scientific outputs.

Open science is really hard, but it's imperative. Africa has different challenges than the United States with open science paywalls. Anytime you go out and search for an article, unless you have a subscription, you can't get it. And that's expensive. So open data allows us to make science happen faster and more equitably. And it includes more view-points. I think about it this way. The National Institutes of Health has an "all of us" Data Research Program, where they're trying to increase the amount of data that is collected from non-European genetics. Be-cause we can tell you everything that happens from folks that are from Europe. But is there a difference when you add—and actually a lot of that is males as well, we don't have enough females, we don't have enough indigenous genetic population information to even compare if there's differences in treatments and drug treatments. That's a big, big deal. So holding this data is very important.

CHAPTER 6

The Shifting Earth

When was the last time you looked at a world map or a globe? When you did look, did anything seem off-kilter or out of place to you? Were all the continents and oceans where you'd expect them to be? Were there any apparent design flaws, or any pieces missing?

If you're like me, you can't look at a representation of the world without noticing that the continents of Africa and South America look like the two halves of a giant cookie that broke in two somewhere along the line. Go look now and you'll see what I mean. Where the deeply carved western coastline of Africa seems almost caved in, the eastern coastline of South America, some 6,122 miles away, bulges awkwardly into the Atlantic Ocean. The two land masses—one concave, the other convex—seem perfectly matched, as though they could spoon together in the bed of the ocean, but only if that 6,122 miles of ocean weren't between them.

If you continue to indulge your imagination, you might see that if it weren't for the intrusion of the five oceans of the world—Atlantic, Pacific, Indian, Arctic, and Antarctic—the subcontinent of India could nestle into Africa's northernmost east coast, with Sri Lanka cozily resting between them, Antarctica sidling up along the lower east coast of Africa, and Australia nuzzling up to both Antarctica and India. Europe, meanwhile, could be seen to balance atop Africa, while Asia would still jut out beyond India, much as it does today.

This is exactly how the earth's seven continents and accompanying land masses are believed to have been laid out some two hundred million years ago, forming one great supercontinent: Pangaea.

Among the earliest to have studied the strangely complementary coastlines of Africa and South America while gazing at a world map (or a globe, no one knows which) was a German polymath named Alfred Wegener. In the early 1900s, Wegener proposed to his fellow scientists that the continents had once been joined together in one vast landmass, but over the course of millions upon millions of years they had broken apart from each other and moved around the globe, propelled by some unknown force. Wegener's improbable theory, which he dubbed "continental drift," didn't catch on in the scientific world, but that didn't stop him from continuing his research. Over time, Wegener found that identical fossil remains of an ancient reptile called Mesosaurus had been found in the southernmost reaches of South America and Africa, indicating that the freshwater Mesosaurus had existed when the two continents were joined together, because they couldn't possibly have crossed the salty ocean. Further evidence came from the snowy climes of Norway, where fossil remains of tropical plants were discovered under the ice and snow, suggesting that Norway wasn't always where Norway is now. Equally curious was the discovery that the complementary coastlines of Africa and South America actually shared identical geological features and that the Appalachian Mountains of the eastern United States were geological doppelgangers of the Caledonian Mountains of Scotland. This onslaught of evidence bolstered Wegener, even if his fellow scientists remained skeptical that the continents could ever move through the earth's crust.

Even with all this evidence to back it up, Wegener's idea floundered, for one simple but essential reason. "[Wegener] didn't have an explanation for why the continents moved," reports the U.S. Geological Survey in its *History of Plate Tectonics*, "just that there was observational evidence that they had."[1]

In 1957, some forty years after Wegener introduced his controversial theory of continental drift and twenty-seven years after his death, his theory was still a hot topic, and on the eve of IGY it was causing a deep schism in the geophysics community. "Continental drift was a huge debate," recalled geophysicist Brent Dalrymple in a 2021 interview. "There was some evidence that the continents had drifted. The shapes between the continents on either side of the Atlantic Ocean looked like they should fit together, and a lot of the geology and paleontology carries right across from one continent to another. It looked like it should've been continuous.

"But at that time, the big barrier to accepting that was the fact that continents have deep roots, and you could not push a continent through the mantle because Earth's mantle is very viscous. Regardless of what kind of force you put on it, it just would not work."[2]

This was because the mantle, which makes up about 84 percent of the earth's mass, is very dense and very deep—1,802 miles deep, to be exact. The mantle has a very important job to do: it separates and isolates the earth's crust from its core and basically holds everything under our feet in place. More or less. So how could the continents break free from this overwhelming geological viscosity?

The debate caused no end of grief for seismologists, glaciologists, and oceanographers. Friendships and research partnerships buckled under the strain; references to continental drift were forcibly edited out of scientific papers. "As an undergraduate," Dalrymple recalled, "I was told that bright young scientists should not work on vague, incorrect ideas like continental drift."[3] Each side even had a name: "Fixists" believed that continents couldn't move, and "Mobilists" believed that they could. Small wonder, then, that the idea that the continents could not and would not push through the mantle remained a firmly entrenched orthodoxy for years, far longer than it should have.

And yet now and then a Fixist would reconsider, soften his or her objections, and begin to embrace the beliefs of the Mobilists. The change never reached stampede proportions, but it was enough to lower the temperature of the debate, just a little.

Untangling the mystery of continental drift was a long, slow, tortuous affair that involved many indefatigable scientists from several scientific disciplines and wasn't really settled until almost a decade after IGY disappeared into the sunset. We've already met Canadian scientist J. Tuzo "Jock" Wilson, the Energizer Bunny of IGY, but we haven't looked into Wilson's work as a geologist and geophysicist. Throughout his work on IGY, Wilson was a firm Fixist and considered the findings of Wegener's work scientifically unsatisfactory, to the point of comparing Wegener's image of continents drifting this way and that to a painting by a certain famous abstract artist: "His view of the earth is like a painting by Picasso in which faces have been turned around or heads displaced relative to bodies. Wegener thought that the surface of the earth had been moved about and displaced by drifting of the continents. . . . No explanation was offered of the cause of these movements, and the idea of drift was not at all acceptable to the proponents of a strong earth."[4]

Wilson's complicated involvement in the debate is woven in and out of the plate tectonics narrative, and the easiest way to trace it is to go through the chronology year by year to see where the strands converge in 1966. It's fascinating to look at the way an idea can change and evolve.

Late 1950s: Before IGY

"Wilson's fantastic memory for details of world geography and geology amazed and even terrified his audience of merely mortal physics and engineering students," recalled the authors of a 2014 tribute to Wilson's legacy. "Tuzo then was a promoter of contraction theory. But, in the late 1950's and following the IGY, he had become very interested in ocean floor geology, and he was rereading all the reports of early explorers such as James Cook. And he was making frequent reference in his lectures to the geology of ocean islands that none of his student audience had ever heard of."[5]

According to University of Missouri–Kansas City professor of philosophy Henry Frankel, a shift occurred in earth science research, brought about by World War II and then the Cold War in the 1950s.

Urgent military needs, Frankel found, drove the development of a whole generation of research tools and instruments, even new fields of science. "Paleomagnetic findings indicated that some continents had changed their positions more or less as Wegener and others had proposed," Frankel noted. "Nonetheless, most Earth scientists continued to reject continental drift, arguing that the results were unreliable.

"The other field that yielded new results was marine geology. In response to the fear of submarine warfare, the United States invested heavily in mapping the ocean floors. Mid-ocean ridges were discovered in all oceans, and explaining their origin became a major problem.

"Harry Hess from Princeton . . . used scientific findings gathered by Scripps Institution of Oceanography and Lamont Geological Observatory researchers as evidence of seafloor spreading. Hess added that his theory of seafloor spreading implied drifting continents."[6]

1960: Two Years after IGY

"By 1960," historian Gordon F. West recorded, "I am sure Tuzo was well aware that contraction tectonic theory was incompatible with what was being learned about ocean basins." West recalled attending a conference in 1960 during which Wilson presented the case for the Fixists, and West was "bemused to see first Wilson and then (the Mobilist defender) display almost identical lists of the global observational evidence they considered crucial, and each to present completely different tectonic models to explain it all. To me, both their models seemed inadequate, although I certainly had nothing better to suggest."[7] "Wilson was sure by August 1960 that ocean floor structure would be the key to a better understanding of continental drift."[8]

Puzzled by the unexpectedly young rocks found in the mid-Atlantic Ridge, Harry Hess proposed in 1960 that new, young mantle material was being pushed to the surface by convection currents, and at the same time old mantle material was being pushed down into sea trenches. Hess called this phenomenon "sea-floor spreading," and it went a long way to explain how continents moved.[9] "If sea-floor spreading was real, the sea-floor adjoining the mid-oceanic ridges ought to show stripes of alternating polarity caused by Earth's alternating magnetic field," wrote geophysicist Ronald E. Doel. "Just such a pattern was recognized two years later in magnetic data recorded by the research ship Eltanin."[10]

In November 1960, *Life* magazine had its own take on the mantle controversy: "The 430-mile level is where the more solid 'skin' of the earth may slide over its inner parts, as though the skin of a grapefruit were to slip over its pulp. Some geophysicists believe the whole upper part of the earth has done this from time to time in the past. They find evidence that the lands and seas which are now at the north and south poles once were in quite different locations. During IGY coal, which is the remains of temperate climate plants, was found in the Antarctic."[11]

The magazine then represented the pioneering work of Alfred Wegener as this "long-discarded theory . . . that continents were once one big mass and then drifted apart. Few scientists today think continents could drift through ocean floors, as Wegener stated." Having said this, the magazine did go on to allow that South America was at that time "slowly moving southward as the east Pacific floor moves northward."[12]

After portraying the earth as a slipping, sliding grapefruit, the magazine then compared it to a second fruit, describing it as a baked apple whose cooling "skin" wrinkles, forming troughs and mountains. The shrinking apple represented the contraction theory, and it fared no better than Wegener's idea. "Modern findings, some of them made during IGY, have almost demolished the contraction theory," claimed the *Life* article. (To continue with the fruit theme as a means of describing the earth, it was discovered by IGY astronomer John A. O'Keefe using gravitational measurements from IGY's many rocket launches that the earth is not perfectly spherical but is in fact shaped like a pear, due to bulging in the southern hemisphere.)

By the time *Life*'s article came out in November 1960, it was also possible to report that "Tuzo Wilson, of the University of Toronto takes a more moderate attitude: that the earth is expanding at the mid-ocean rift zones at about a fiftieth of an inch a year."[13]

1961: Three Years after IGY

In his book *IGY: The Year of the New Moons*, Wilson wrote, "Because New Zealand is an isolated island, it is an obvious place to look for evidence on continental drift. At the present time we not only are not sure about the existence or rate of these motions but we also do not know whether the continents are growing. The evidence, I believe, suggests they are."[14]

"It was clear to us in the Geophysics Laboratory at University of Toronto that Wilson was keenly excited to be involved in rethinking global tectonics, but we had no idea where this was going to lead," recalled West.[15]

1963–1965: Six+ Years after IGY

By 1963, Wilson's thinking on continental drift had evolved dramatically. Why? Because he now had at his disposal a year and a half's worth of data collected through his own work and that of his IGY colleagues, and because, good scientist that he was, he was capable of changing his mind when presented with convincing data contrary to his own findings. Interestingly, though the increasing amount of evidence for continental drift continued to impress Wilson, there was still an issue the Mobilists couldn't address. The theory hadn't explained why active volcanoes could be found thousands of miles from any known plate boundary, but Wilson was able to explain this as the action of plates sliding over fixed "hotspots" in the mantle from time to time. This, Wilson theorized, was the force that created the Hawai'ian islands.

Wilson followed up in 1965 with the discovery of "transform faults," sometimes referred to as conservative plate boundaries. A transform fault involves two plates sliding and scraping alongside each other in a horizontal motion, rather than smashing into each other, causing one plate to go up and one plate to go down. This sliding action can be seen every time there is a seismic event involving the San Andreas fault, when the North American and Pacific plates rub up against each other. At the time of Wilson's discovery, transform faults were regarded as the missing piece in the puzzle of plate tectonic theory.

"Transform faults, which the Canadian geophysicist Tuzo Wilson predicted in 1963 . . . provided a physical explanation of the movement of crustal blocks past one another along plate margins," Doel wrote. "Equally important was the determination, through studies of continental and oceanic earthquake data, of the outlines of the major 'plates,' seven in all, each built around continental landmasses."[16]

1966: Nine Years after IGY

"In 1966, seafloor spreading was confirmed, and plate tectonics, the modern version of continental drift, was accepted soon after. According to plate tectonics, the Earth's surface is divided into a dozen or so major plates that move relative to each other."[17] "I became a convert to continental drift and sea-floor spreading one day in late spring 1966, when I obtained my first mechanism solutions of earthquakes along the Mid-Atlantic Ridge," explained seismologist Lynn Sykes in a 2019 interview. "They agreed with Tuzo Wilson's hypothesis of transform faulting along huge fracture zones that displace segments of ridges. My finding showed that the Mid-Atlantic Ridge was growing along its ridge crests and that continents on the two sides of the Atlantic were moving apart."[18]

"Back at the Geophysics Lab, we were all busy with our own local projects," recalled West. "It was more than a decade later when we eventually learned that Larry Morley, one of Wilson's early PhD graduates and by then Director of the Geophysics Division of the Geological Survey of Canada, had understood the full story of magnetic imprinting at ridges, but was rebuffed on two attempts in early 1963 to publish it, so no one became aware of it until much later."[19]

1967: Ten Years after IGY

"By late 1967 the core concepts of modern plate tectonics theory coalesced within the North American geological community," wrote Doel. "The authority of plate tectonics as the reigning interpretation of Earth's structure is not simply its theoretical elegance but its predictive ability."[20] In 1994, thirty-six years after IGY, Robert Dietz of the U.S. Coast and Geodetic Survey wrapped up the history of plate tectonics with these eight words: "It was an idea whose time had come."[21]

And so, despite decades of controversy, false starts, and dead ends, IGY, Tuzo Wilson, and a vocal, determined cohort of geophysicists and geomagneticists finally paved the way for an entirely new view of the planet on which we live. Science could finally agree that Wegener was close to being correct, that the continents did indeed move, ever so slowly. Perhaps just as important, science could finally agree on the force that made this movement occur.

"The outer 100 miles of the earth is composed of about 15 plates of strong, rigid rock that move with respect to one another, much like cakes of ice on a river," said Sykes in her interview. "Most earth deformation, earthquakes and volcanoes occur at the boundaries of plates, where they either move apart from one another as along the Mid-Atlantic Ridge, converge as along island arcs and deep sea trenches like Japan, or slide by one another as along California's San Andreas fault."[22]

"When I was a student in undergraduate school and graduate school . . . the oceans were considered the oldest features on the earth. And they were going to be a source of great mineral wealth," recalled Brent Dalrymple in a 2021 oral history interview conducted by the American Institute of Physics. "And after plate tectonics, we learned that, in fact, they were the youngest features on the earth, aside from individual volcanoes and features like that, and that they were not a great source of mineral wealth."[23]

Dalrymple's recollections illustrate how a scientific advance like the discovery of plate tectonics can thoroughly change our understanding of how the world works. "We didn't know why certain things existed," he said. "Like the Appalachian Mountains, for example. Why were they there? Well, there were all kinds of goofy explanations, but it just turns out that that's where two continents banged together over 400 million years ago. And that's something that plate tectonics told us."[24]

"Alfred Wegener proposed continental drift back in the 1920s, and a lot of people think that scientists ignored it," Dalrymple said. "But my experience was just the opposite, that it was debated very heavily, at least at Berkeley, where I went to graduate school. You could see the evidence for it, and you could also understand the physics that precluded it. And not until plate tectonics came along did the solution become obvious, that is, that the continents were not plowing their way through the mantle but riding on thick lithosphere plates that were sliding over the deep mantle."[25]

When asked what kinds of technological advances made the plate tectonic revolution possible, Dalrymple replied that increased IGY-sponsored ocean exploration played a big role. "Some of it was increased exploration at sea. And outfits like NOAA towing magnetometers around without really knowing why they were doing it, but making these beautifully colored magnetic maps of the sea floor. Measuring the ages of rocks and minerals required the development of equipment to measure isotopes very precisely. And most of those revolutions came about, like, in the 50s and early 60s. And so, part of what led to the plate tectonics revolution was the ability to measure things that needed to be measured in order to solve certain problems in geophysics and geology."[26]

"There were magnetic stripes on the seafloor that were called positive and 'negative' magnetism," Dalrymple went on. "The pattern that the magnetic stripes made on the ocean floor mapped right onto our geomagnetic time scale. So, these stripes were essentially the result of seafloor spreading. Hot material came up from the mantle and created new ocean crust, and as that new crust cooled, it recorded the magnetic field at the time. And what you had was a conveyer belt that was moving out, forming new magnetic stripes on either side of the ocean ridges. So, our geomagnetic time scale turned out to be one of the keys to plate tectonics, and the proof of seafloor spreading. The geomagnetic time scale has often been called the Rosetta Stone of the theory of plate tectonics."[27]

"It solved the problem of continental drift," Dalrymple said. "What's happening is, there are lithospheric plates, which include both mantle and crust, and these plates are a couple hundred kilometers thick. The plates are essentially sliding over a more amenable zone in the mantle. And so, the continents are really just riding on top of the plates. And that solved the big problem. It showed that continental drift did happen in a physically realistic way."[28]

Or, as IGY geologist Charles Drake decisively put it, "If [the seafloor] spread, the continents had to move."[29]

*   *   *

More about the earth's crust: The discovery of plate tectonics certainly was a marvel, but I'd like to end this chapter with a brief summary of the objectives and results from some of the lesser-known, smaller surprises of the IGY seismology program, taken from the 1965 *Report on the U.S. Program for the International Geophysical Year*:

Summary of Seismology Program
The seismological program consisted of: a) earthquake studies with emphasis on special studies of long period waves; b) seismic exploration of the crustal structure; c) microseismic studies; d) the extension of earthquake studies and seismic exploration techniques to Antarctica. Projects provided for the procurement, installation, and operation of seismic equipment at several new stations and for seismic surveys

on oversnow traverses. Other projects were established for the collection, reduction, analysis, and publication of data.

Reduction of Seismic Records on Antarctic Traverses

Objectives: This project at the University of Wisconsin Department of Geology provided for the reduction and publication of reflection seismic data gathered during the antarctic oversnow traverses carried out from Little America, Byrd and Ellsworth stations during the antarctic summers of 1957–58 and 1958–59. The project included transcription and reduction of data recorded in the field, measurement of arrival times, plotting of travel times and appropriate graphs, and corollary determinations of velocity and depth of reflecting surfaces. The data collected during individual readings had to be coordinated with the entire series, taking into account internal consistency as well as gravity values obtained.

Results: Seven major oversnow traverses, supplemented by several shorter trips and one airborne traverse, covered approximately 12,000 km of track in Antarctica. The primary seismic measurements at locations about 30 miles apart on each traverse were ice thickness and depth of water beneath the ice for those stations on ice shelves. The travel time of the compressional echo served to locate the rock surface, multiple travel paths or shear waves were used on the ice shelves to determine thickness of floating ice. In addition, many short refraction profiles were shot to give detailed determinations of velocity variations in near-surface snow and ice. Long refraction profiles were shot to obtain information on seismic velocities throughout the icecap and in the underlying rock, using both vertically and horizontally oriented geophones to record all possible phases.

Sea Exploration, Atlantic Ocean

Objectives: Under this project at Lamont Geological Observatory, seismic refraction and reflection measurements were made on a traverse along the eastern and western basins of the Atlantic Ocean, in coordination with vessels from the Woods Hole Oceanographic Installation and the National Institution of Oceanography (England) and the Argentine Navy. In addition, a seismometer was designed for emplacement and recording on the ocean floor (where the signal to noise ratio was expected to be greatly enhanced).

Results: Although most of the measurements in the Caribbean Sea were made prior to the IGY, the results are of particular interest. Geographically and in terms of depth of water, it has not been clear whether such areas should be regarded as part of the continent or of the oceans, or whether they represent a transition phase between the two.

Seismic Sea Exploration of the Southeast Pacific

Objectives: This project at the Scripps Institution of Oceanography provided for two-ship seismic refraction studies of areas of geological interest. The measuring program had as its objectives the exploration of the structure of the earth's crust down to the Mohorovicic Discontinuity. Because this region has been largely unexplored by modern geophysical methods, one of the paramount questions has been: are there any extensive regions whose crusts differ from the "typical" Pacific crustal structure? One such possible region is the elevated area called the Easter Island rise on which indications of anomalous structure were found on the CAPRICORN Expedition.

Results: In general, the oceanic measurements corroborated earlier seismic studies in the North and Central Pacific. As a rule the sedimentary layer is only a few hundred meters thick except in the equatorial barbonate area where the thickness is about two times the normal oceanic value. There is perhaps more variation in overall crustal thickness than observed previously, and although the thickness of the basic crustal layer having a velocity of about 6.7 km/sec is fairly constant, many stations showed an intervening intermediate velocity layer which averages about 1 km in thickness and has a velocity varying from 4 to 6 km/sec. Unusual results were obtained in four areas.

a. In a deepwater embayment in the Tuamotu Island platform, normal oceanic structure is found although the surrounding platform is a shallow water area.

b. Just east of the Easter Island Rise, the crustal layer appears to be only about 4 km thick. Conditions are also different in that the thickness of the upper crustal layer, about 2 km, equals that of the lower layer, which has a velocity of about 6.7 km/sec.

c. Under the Nasca Ridge, beneath 2900 m of water, the crust appears to be about 15 km thick.

d. At four locations on the Easter Island Rise, despite unusually long profiles, no velocity greater than 7.5 km/sec is observed. This is also an area where the heat flow is observed to be about five times the normal value. As other parts of the Easter Island Rise appear to be characterized by normal crustal conditions and normal heat flow, this correlation strongly suggests that the mantle velocity may be affected by abnormal thermal conditions.

Interview with Spencer Weart, PhD

Former Director of the Center for History of Physics of the
American Institute of Physics (AIP)

Mark O'Connell: What are some of the lasting benefits of IGY?

Spencer Weart: The Keeling Curve: a curve showing the rise of the carbon dioxide atmosphere. What do you think such a curve looks like?

M.O.: A hockey stick?

S.W.: No, sorry. Well, actually, it would if you go far enough back. Charles David Keeling started measuring carbon dioxide in the atmosphere in 1956. So it just goes up and climbing a little faster every year. Roger Revelle was a great oceanographer who had realized, and this was basically the thing that kicked the whole thing off, he realized that the oceans can't absorb carbon dioxide as fast as [it] is being emitted up until 1956. Everybody thought that, well, sure we're emitting a lot of carbon dioxide, that all dissolved into the oceans. And Revelle said, no, no, we're emitting it too fast for the oceans to take up. So it's going to rise in the atmosphere, and that'll bring global warming.

Keeling had measured carbon dioxide in the atmosphere and got a rough idea what the present level was, but just kind of a rough idea. So Revelle said, well, I'd like to see if my idea's correct. Let's measure carbon dioxide all around the world now. And then we'll come back in twenty years and see if it's going up.

Where are we going to get the money to do that? Nobody wants to do such a funny, weird thing like that. And the answer was the IGY. The IGY had suddenly appeared with a big pot of money. So Revelle and colleagues went to Harry Wexler, who was the go-to guy at the Meteorological Bureau, and said, "Hey, we want some money to measure carbon dioxide in the atmosphere. It might have some long-term implications, you know, might actually mean something for civilization." Because we were well aware that if you really did double the carbon dioxide in the atmosphere, then you know, you'd be able to navigate the Arctic Ocean and there'd be great droughts here or there. And so they thought it was an interesting scientific question. And Wexler says sure, okay, here's $20,000. Go ahead and do it.

Keeling was a weird guy. Keeling only ever wanted to do one thing in his life, and that was to measure things precisely. And when he set on carbon dioxide for his whole life, all he ever did was measure carbon dioxide precisely. And then he got the actual base level very precisely. And in two years in Antarctica, he found that the level in the atmosphere was rising. Didn't have to wait twenty years; he saw it in two years.

And that essentially got the ball rolling. From then on, it was a serious scientific question that was worth investigating. You know, given that Keeling says the carbon dioxide level is rising, and given that Keeling is a fanatic who never gets anything wrong and those kinds of things, and given that rising carbon dioxide is bound to cause global heating, then, you know, we could be in deep shit and let's look into it.

M.O.: What other IGY successes can inspire us today?

S.W.: The first satellites were done under IGY, and satellites, of course, had become absolutely essential. How do we know the temperature and the winds and everything else at every point on the earth's surface? It's thanks to satellites. Until they put satellites up, nobody had any idea, for example, about how cloudy the Southern Hemisphere was. Obviously, you want to do global climate models, you need to know where the clouds are in the Southern Hemisphere.

And we have to remember that the people who founded the IGY were aiming for exactly this kind of thing. The whole purpose of the IGY was to jump-start international scientific developments. And if you go to do climate, there is nothing more international. I mean, all sciences are international, but it's only climate scientists who need permission to enter twenty different ports around the world if they're going to do oceanography.

M.O.: Did you say ports permission?

S.W.: If you're an oceanographic research vessel, you have to go into port from time to time. The first great oceanographic vessel, which was called the *Challenger*, was put out by the British, and they could put in all around the world at British ports because that was the British Empire. Anybody else had to get permission to take their oceanographic vessel into a port, so there's a lot of diplomacy, and a lot of international cooperation is just essential. Obviously, if you want to know the weather in New York, now you need to know the weather in Canada two days ago. So the Canadians have to be nice and tell you. Okay, this can be a serious problem. If, for example, Germany's at war with Great Britain, and Germany wants to know what the weather is going to be in Normandy, they have to rely on two or three submarines out [in] the middle of the Atlantic. Okay, the submarines see a storm, and then the Germans say, "Okay, on the sixth of June is going to be very stormy." On the other hand, the British have a lot more ships, and they have Greenland and Iceland, and so on. And they say, "Well, yeah, there's going to be a break in the storm. We can land on the sixth of June." Which they did, while the German generals were off on vacation, assuming that nothing would happen. So it's very international. And you really need to have nations talking to one another, if you're going to deal with climate.

M.O.: You've touched on efforts to measure changing levels of carbon dioxide in the atmosphere. Didn't some of that work involve some very deep drilling into the Antarctic ice?

S.W.: Does the word *Vostok* mean anything to you? Okay, this is where the Russians went to drill into the ice because it was sort of the highest and deepest place; it's two miles deep. And they brought up this two-mile core and this was the mid-1990s. They got both the temperature and the carbon dioxide very accurately, all the way through the past Ice Age. Eventually they went through the past four ice ages and one thousand years of record. And they saw that carbon dioxide and temperature just went up and down together, they were just so obviously clearly linked. And then you can do numbers.

Well, Vostok got the numbers, and they said, "Oh, if you double the carbon dioxide, temperature goes up three degrees." Now, this is sort of a basic feature of the philosophy of science. If you use two totally different methods, and you have the same numbers, that makes you feel that you have touched reality. Okay, so this is what the last type results did. It gave that second method of getting the same result. And I think it was from that point on that, you know, scientific research, scientific opinion had been moving in this direction, but that gave a big jump to the conviction that global warming was real.

M.O.: Is science moving in the right direction now?

S.W.: I know it is happening. I know what is happening. At this moment ten thousand scientists are working within a framework set by the Intergovernmental Panel on Climate Change (IPCC) and the world climate research program coordinating their research, largely as volunteers. The climate science community is an enormous engine of knowledge production coordinated entirely by the scientists.

Yeah, I mean, look at the IPCC. The United Nations helps administer it. It's what it says on the labels—intergovernmental. It doesn't answer to the Secretary General, doesn't answer to the General Assembly. It answers to representatives appointed by the world's governments, nearly all of whom are scientists. If you're Burkina Faso, you don't have a climate scientist. You appoint somebody from your meteorological bureau, but they're all scientists.

M.O.: I appreciate you spending this time educating me and my readers.

S.W.: You know, I'll tell you my attitude toward this. I was supported by society, for most of my life to have fun by researching stuff. And I have an obligation [to] society to tell them what I learned.

CHAPTER 7

Rockets and Satellites

Launching an artificial satellite into earth orbit was an integral element of IGY almost from the start, and a rather spectacular one at that. As early as 1954, the use of man-made satellites to explore the high atmosphere was considered an IGY must. Given the fact that the nearly seventy German V-2 rockets and 120 rocket scientists that found their way to the United States in the aftermath of World War II had given American scientists a crash course in rocketry, it was an easy and obvious call, even if the physics of the time seemed impossible. Due to the proliferation of science fiction words and images in novels, movies, television shows, and even comics, Americans in the mid-1950s were falling in love with the idea of space travel. And it wasn't just kids with Buck Rogers toy ray guns making space travel more mainstream than ever; in the 1950s, the very serious *Collier's* magazine published a series of enormously popular illustrated articles written by scientists describing how Americans might someday journey to the moon and then on to Mars. Some practical scientists even laid out plans for the construction of an orbital space station, a "wheel in space" that would create its own artificial gravity and function as a wayside for astronauts headed out to explore the solar system. In this social moment, the government's nascent efforts to launch satellites to explore Earth's upper atmosphere and then the void high above it only whetted the public's appetite for taking even one small step off planet Earth.

There was no guarantee, however, that those ambitions could ever be satisfied. Undeterred, the U.S. IGY committee formed the Long Playing Rocket (LPR) committee in early 1955 under the direction of Dr. Fred Whipple of the Harvard-Smithsonian Observatory.* The committee's remit was to study and report on the geophysical possibilities of orbital flight. Thus, the LPR committee took on the responsibility of determining the "technical feasibility" of an orbital satellite. The committee would report to the IGY board on "desired orbit . . . controls, motor, manpower, timing, budgets, cost estimates," and other practical matters.

In a matter of months, the LPR committee was able to present an initial proposal to the executive committee: this was to launch into orbit a ten-pound artificial satellite of spherical shape with a twenty-inch diameter. The satellite, dubbed Orbiter, would be painted white or some other highly reflective hue, so that it could be observed from the ground at twilight using binoculars or an Askania cine camera. The LPR committee also called for the creation of a global tracking network to maintain visual contact with the satellite in the course of its orbit. Finally, it was proposed that payloads for the satellite be designed to address a variety of scientific concerns, including

- precise geodetic measurements
- determination of upper air densities
- measurements of solar radiation
- measurements of particle radiation
- determination of current flows in the ionosphere associated with magnetic storms and radio blackouts
- determination of hydrogen in interplanetary space

The LPR committee estimated that this mission could be accomplished in two to three years, and in short order the IGY committee approved a $10 million budget for the project. Then, on July 29,

---

* Some readers may remember listening to music on "long-playing records," vinyl discs with music recorded on both sides. The name Long Playing Rocket was a play on this idea, which made sense at the time.

1955, President Dwight Eisenhower made it official: "On behalf of the President," his press secretary, James Hagerty, announced in a press release, "I am now announcing that the President has approved plans by this country for going ahead with the launching of small un-manned earth-circling satellites as part of the United States participa-tion in the International Geophysical Year which takes place between July 1957 and December 1958. This program will for the first time in history enable scientists throughout the world to make sustained observations in the regions beyond the earth's atmosphere.

"The President expressed personal gratification that the American program will provide scientists of all nations the important and unique opportunity for the advancement of science."[1]

When the project became official, that's when difficulties arose. Those former German V-2 rockets that had found their way to the United States had been claimed by and distributed to several U.S. gov-ernment agencies to guide them in development of their own bigger, better, faster, stronger rockets and missiles, some built for scientific research, others for military and intelligence purposes. So once the commitment was made by the president to place a ten-pound, twenty-inch-diameter shiny globe into earth orbit by the end of 1958, the scientists and the military men had to decide whose rocket (or missile) would get the honor of launching that globe into orbit.

First in line was a team of scientists who were developing Project Orbiter with the support of both the U.S. Army and the Office of Naval Research. This proposal was boosted by former Nazi rocketry expert and future NASA head Dr. Werner von Braun, who proposed using the Redstone missiles he was developing at a facility in Hunts-ville, Alabama, as the launch body. The Orbiter proposal would place a six-pound sphere into orbit and would track that sphere with the as-sistance of the U.S. Army and the Harvard-Smithsonian Observatory.

Next came the U.S. Navy's Naval Research Laboratory, which had been developing a rocket program of its own utilizing a Viking rocket, even as the Office of Naval Research was helping develop Orbiter. This effort would use a twenty-pound, thirty-inch-diameter satellite lifted to an orbit of 150 miles. Not wishing to be left out, the U.S. Air Force was proposing an orbital launch based on an intercontinental ballistic missile (ICBM) called Atlas.

So by March 1955, the U.S. IGY committee found itself with three very similar and very capable launch programs that could be ready to launch:

- the Redstone-backed Project Orbiter
- the Naval Research Laboratory's Viking project (later renamed Vanguard)
- the U.S. Air Force's ICBM-based Atlas project

With competition and confusion rising between government agencies and scientists, the secretary of defense needed to send a signal and move ahead on one of the three projects. But there was yet another hurdle: whichever project was chosen, its development could not interfere in any way with the development of the ICBM arsenal. Of the five-member advisory board recruited to choose the best of the three proposals, three chose the NRL Viking/Vanguard plan and two chose the Orbiter. They all steered clear of the Atlas plan, fearing that even in a best-case scenario it could create bottlenecks with the production of ICBMs. In the end, the safer Vanguard project got the nod.

Ironically, the mammoth ICBM operation then created bottlenecks for the Vanguard project. Cold War fears put increasing pressure on the military to ramp up the ICBM program, meaning that "when it needed parts, help or money, Vanguard had to go to the end of the line."[2] It didn't help that many military figures saw the satellite project as "a neat trick" that had no long-term strategic value.

Those critics may have been expressing anxieties about Vanguard funding coming at the expense of their own programs, however. Bear in mind that every one of these decisions concerning the fate of expensive government programs and technologies can decide the fates of hundreds of careers and the rise and fall of aerospace companies that may be here today but gone tomorrow when passed over for a government contract or when a lucrative contract is abruptly canceled. When Vanguard is selected as the IGY rocket mission, the fortunes of anyone associated with its rival programs can be ruined in the blink of an eye, with whole careers rendered obsolete. That is, unless they can transfer their assets, technologies, and resources to a different government contract.

Once the matter of which launch vehicle would play a starring role in America's IGY space program was decided, another issue arose, this time with far more dramatic repercussions. Although the Soviet Union had been among the earliest nations to formally commit to participating in IGY, it had a seemingly insurmountable conflict where the event was concerned. Recall that all participating nations agreed to share their IGY-derived research data freely with other scientists in other nations, and in exchange they would have access to the work of other IGY scientists and researchers. That may have been fine with some of the Soviet scientists, but not so fine with many Russian rocketry experts, as well as the government agencies that funded and monitored their work.

It was known in the United States that the Soviets were making use of the former Nazi scientists and rockets that they had acquired at the end of the war in much the same way the United States was doing. In some ways, the Soviets were keeping pace with America's rocket development, in some ways they were falling behind, but it was the areas in which the Soviets were ahead that caused concern. The Soviets were committed to share all their data with all the IGY countries, but they were loath to share sensitive information about the newest developments in their rocketry program with any nation, least of all with their archrival, the United States.

"Launching" nations such as the United States and the Soviet Union were asked to provide the following data:

- launching sites of both rockets and satellites
- rocket firing schedules

- the approximate period of satellite launchings to enable ground stations to make the necessary preparations
- announcement of each satellite launching within an hour, with a report on its success within three hours

IGY participants were expected to share the orbital data from a rocket launch within five months, and a "complete tabulation of the reduced, calibrated, and corrected data radioed from the satellite was to be made public in eight months." In addition, the nation launching the vehicle would make its data accessible to other scientists from other nations "for assurance against misinterpretations."[3]

The Soviet scientists were not fully agreeable, to say the least. The dynamic is best illustrated by a phenomenon observed at an IGY conference held in July 1957, as recounted by Walter Sullivan: "During this week-long conference the American scientists, as tactfully as they could, made their Russian colleagues aware of their disappointment at the lack of communication which the Russians had produced on their satellite program. The Russians, in turn, made it clear that they found it unseemly for scientists to 'boast' about their experiments until they were done."[4]

As a result, Sullivan noted, "Few in the West were aware of the swift Soviet advances in science and technology that had taken place in the four or five years before the launching of the first Sputnik."[5]

One significant area of tension had to do with tracking satellites once they were in orbit, a particularly challenging task. On learning that the Soviets would be using a different radio frequency to communicate with their satellites than that used by the United States, Americans complained that the Soviets' technology was incompatible with that of the United States. Further, the Soviets protested some of the IGY data reporting rules, most noticeably one that required any launch nation to submit the location of every launch site used for IGY. In an additional request, the Soviet Union wanted to only have to announce successful launches and to increase the reporting deadline for such launches from one hour to two. By and large, the changes seemed to favor the Soviets—who, the United States eventually learned, had their share of launchpad disasters—and created an impression that they were concealing valuable information.

For a participating IGY nation to lobby for fewer and less stringent data reporting requirements—and to do so to put a rival nation at a disadvantage—created a certain amount of discomfort among member nations, but that was nothing compared to what was to come.

The Soviet Union dazzled and terrified the world when it launched Sputnik, the first artificial satellite, into space. On October 4, 1957, just a few months into IGY, a Soviet launch vehicle delivered a shiny metal ball to earth orbit, to the surprise and shock of most of the population of Earth. At that moment, one of the most significant accomplishments of IGY was introduced to a nervous world: the inauguration of the space race. Faced with paradigm-shattering proof that earth orbit of a man-made object was no longer an impossible dream, IGY directors had little choice but to recognize the accomplishment of the Soviet rocket scientists and let bygones be bygones where the squabbling over following the rules and maintaining the spirit of cooperation embodied by IGY was concerned. This was big.

While millions of people in the United States sank into a deep funk at the realization that the Soviets had put them to shame with Sputnik, a few scientists rallied and began to prepare a proper response.

Like much of the world, many American scientists had assumed that the American Vanguard would beat the Soviets into space, and they had been building a global tracking network to keep an eye on the Soviet satellite while it was—in theory, at least—keeping an eye on us. Dr. Fred Whipple, director of the Harvard-Smithsonian Observatory, tasked his junior colleague, Dr. J. Allen Hynek, with devising an astronomical undertaking that had never been attempted before: creating a way to locate an artificial satellite in the night sky and then track its progress and ultimately determine the path of its orbit.

Hynek and Whipple devised a two-pronged approach. First, Hynek would recruit thousands of amateur astronomy hobbyists called "Moonwatchers" to form up teams of volunteers to watch the night sky. Then, when and if one of these teams spotted something in the sky that didn't belong there—basically, anything that visibly moved—they would pass the information on to one of twelve special observatories situated at high altitude sites around the globe. Astronomers at those twelve locations would then "look" where the Moonwatchers had indicated and record the passing of the satellite with the use of twelve-foot, purpose-built astronomical cameras. That was prong number two.

"Moonwatch was more than fulfilling the expectations of its creators and was demonstrating its ability to provide data of singular scientific significance," reported E. Nelson Hayes of the Harvard-Smithsonian Observatory. "Hundreds of people of widely differing personalities and vocations had responded to the romantic and even adventuresome appeal of Moonwatch."[6]

The satellite tracking project unleashed a bit of pioneer spirit in its leaders and participants, as everything they were discussing, planning, and doing was something that had never been done before, or in many cases never been attempted before. The twelve Baker-Nunn astronomical tracking cameras scattered across the globe are a case in point.

"The manufacture of the cameras was one of the finest achievements of American industry," Hayes wrote. "Of entirely new design and of complex structure, 12 of them had to be built without the construction of a prototype, and without the testing of the individual components of the system. The cameras were built almost concurrently, and the first one completed had to work. And once the large components were put into production, there was no opportunity to change any of the details."[7]

As plans for tracking the Vanguard satellite developed and became more complex, it became clear that the kind of scientific data collected by the Moonwatch volunteers and the professional astronomers could dramatically increase our knowledge and understanding of

- the effects on the earth and the ionosphere of solar ultraviolet light, cosmic and solar X-rays, and other radiation;

- the physics of the upper atmosphere as it relates to more accurate long- and short-range weather forecasting;
- the points in the upper atmosphere at which energy is either absorbed or radiated, and the problem of energy balance and dynamics of the upper atmosphere;
- the disturbances in the atmosphere that result from solar flares and solar radiation;
- the relationship between conditions in the upper atmosphere and the weather at lower levels;
- the variations in density and temperature at different levels of the upper atmosphere;
- the nature and cause of the aurora;
- the forces that produce the changes and fluctuations in the earth's magnetic field;
- the variations in composition and thickness of the earth's crust;
- the exact size and shape of the earth; and
- the sizes and relative positions of the land masses of the earth.

Interestingly, Drs. Whipple and Hynek concerned themselves with what type of person they could recruit to operate the remote camera stations as much as how well the cameras themselves might work. "What was needed was a person who had eagerness, enthusiasm, a spirit of adventure, and especially a sense of responsibility," wrote Hayes. "The first observers were essentially romantics, men who had . . . a common interest in this new age of satellites and intense curiosity about science and the world."[8]

Of course, all of those romantics scanning the heavens, watching for movement in the night sky—amateurs and professionals alike, all around the globe—assumed that they would be playing a role in tracking the first Vanguard satellite. It must have come as a surprise for them to realize that all their efforts were being put to use to track a Soviet satellite instead, a Soviet satellite that had just turned the world of science upside down.

The United States reacted quickly to the bombshell news, of course, announcing that it would soon respond with the first Vanguard launch. But it was not to be. The first Vanguard wobbled and burst into rocket-fuel flames only four feet after leaving the launchpad, and it was all shown, humiliatingly, on national TV. The next Vanguard launch attempt also failed when the rocket broke in two and crashed back to earth sixty seconds after liftoff. If it was the Soviets' mission to crush Americans' souls, they could hardly have been more fortunate.

"It was a shocker," recalled Herbert Friedman of the Naval Research Laboratory. "We had been having a meeting all week, over at the National Academy—an IGY meeting with the Russians—in which we were describing our two programs, and trying to get the Russians to accept frequencies which we had chosen for the Vanguard program, which they did not agree to do. And the debate went on, in the sense that we were ready to launch and they were not. And yet they were giving us signals that we just didn't appreciate. They were about ready to go. By the end of the week of that meeting—I think it was a Friday night—the news came through that they had succeeded in launching Sputnik, and it really took us completely by surprise. We could look back on that week afterwards and realize that they were telling us that they were about ready to go, and we just were not reading the message."[9]

*   *   *

Even worse than this blow to U.S. pride was the new fear that the Soviet Union was now capable of attacking the United States with nuclear missiles from the sky, with little or no warning. For all any American knew, Sputnik itself could be the first salvo in a surprise attack by weapons launched from Siberia and flown over the North Pole to descend on vulnerable American targets. After all, didn't that strange contraption pass overhead every ninety minutes, and didn't it emit an unnerving "beep" sound every time it passed?

By design, Dr. Hynek's tracking project worked flawlessly, but by pure chance, he was the only astronomer on duty at the Harvard-Smithsonian Observatory the night Sputnik was launched, and so it fell to him to respond to the hundreds of phone calls and telegrams from reporters wanting to confirm the news that the Soviets could now launch a second Pearl Harbor attack at a moment's notice. As a result, Hynek's voice became the voice of reassurance to millions of frightened Americans.

But there were secrets. "After the launching of Sputnik I, the Russians were notably incommunicative about the final stage rocket that placed it in orbit," Sullivan wrote. "Everyone recording satellite signals was to send them to the launching country, which would ultimately publish the results. Moscow was not obliged, under IGY rules, to make public the key to the signals, but this was not generally understood in the West and, furthermore, the dearth of other information from Moscow fostered a widespread belief, at least in the United States, that the Russians were not living up to their IGY obligations."[10]

So in the end, the dawning of the space race brought both good and bad: IGY's crowning glory and the world's worst nightmare.

* * *

More about rockets and satellites: Although Sputnik stole the show in 1957, there was a great deal of lesser-known work done to advance and support humankind's transition from earth flight to space flight, as shown by the 1965 *Report on the U.S. Program for the International Geophysical Year*:

Summary of Antarctica Activity Program

The specific objectives of the U.S. earth satellite program were

- To place an object in orbit and prove this by observations.
- To obtain a precision optical track for geodetic and high-altitude atmospheric drag study purposes.
- To perform experiments with internal instrumentation.

Establishment and Operation of Optical Tracking Stations
Objectives: These projects at SAO provided for the establishment of twelve satellite tracking stations throughout the world. The reduced data obtained from the optical stations would permit the calculation of definitive orbits for use in correlating with on-board and ground-based experiments and provide valuable data for scientific research. In addition, quick-look, or field-reduced data would be valuable for certain geophysical research by satellites, as well as serving as the basis for continued predictions for the tracking network itself.

Results: The precision satellite tracking cameras performed as expected and produced a collection of a large number of observations. For example, during the month of April, 1959, 126 observations were made at Woomera, Australia, at this station, 414 observations were made from the start of observations to May 1959. The average monthly number of observations, as of May 1959, ranged from 52 at Woomera to 16 at Shiraz, Iran and Jupiter, Florida. Variations in the geometrical positions of the satellite orbits and in local meteorological conditions undoubtedly accounted for the wide range. The individual observations were collected at SAO and formed the basis for the computation of orbital elements, from which station predictions could readily be derived, and for precision orbits used in basic research such as the derivation of atmospheric properties and other geophysical results.

Visual Observing Program—Moonwatch
Objectives: This project, organized by SAO, was designed to provide satellite sighting information in the early phase of a satellite's lifetime in order to obtain a preliminary ephemeris for the use of the precision optical network. In addition, the visual observers could hope to re-acquire "lost" satellites and to secure information in the late phases of a satellite orbit, as the satellite approached nearer to earth and its orbit became erratic. Finally, the visual program complemented the radio tracking networks for many satellites and served as the major preliminary tracking service for

satellites whose telemetry failed and for parts of the rocket which stayed in orbit.

Results: As of June, 1959, approximately 8,000 volunteers had participated in Moonwatch, comprising a total of some 250 teams, of which 193 supplied observations to Smithsonian Astrophysical Observatory (SAO). Achievement certificates were recommended for 50 teams and awards to 205 individual participants for meritorious contributions to the program. The teams were located in North and South America, Africa, Europe, Asia, the Middle East, as well as in the Arctic Basin on Station C and Ice Island T-3. Moonwatch teams were alerted for each U.S. satellite launch, particularly for experiments such as the inflatable sphere for which no radio tracking was possible of some of the elements placed into orbit. From the beginning of the Moonwatch operation in October, 1957, through June, 1959, nearly 10,000 observations were sent to SAO. The Moonwatch program continued after the IGY period and teams are still contributing importantly to the satellite tracking program.

As an example of the function of Moonwatch during special situations, a search plan was initiated during 1959 for Satellite 1958 Beta 1 (the non-broadcasting element of Vanguard I) which had been "lost." Using an observation made May 6 by the two Albuquerque, N.M. teams at a joint session, Arthur S. Leonard, leader of a Sacramento, California team, modified the orbital elements of 1958 Beta 2 (Vanguard I); using this prediction, his team observed Beta 1 on May 10, allowing for further revisions to the orbital predictions. Several California stations as well as the Albuquerque station sighted Beta 1 the next evening. These Moonwatch sightings allowed orbital predictions to be improved so that precision optical observations began on May 12.

Cosmic-Ray Observations

Objectives: This project at the Department of Physics, State University of Iowa, was designed to make comprehensive measurements of the total intensity of cosmic radiation as a function of time, latitude, longitude, and altitude. As a result of the first experiment, when it was discovered that at a certain altitude the radiation level increased sharply to far beyond normal cosmic-ray levels, the program was reoriented to investigate in detail the structure of this zone of intense radiation.

Results: The orbit of Explorer I (1958 Alpha) was eccentric, with a perigee of about 360 km and an apogee of about 2,500 km, the inclination of the orbit was 33.5°. For much of the time, the counting rates seemed to be what was expected on the basis of extrapolating plots of intensity vs. altitude, as determined at lower altitudes with balloons and rockets. However, on some of the passes the counting rate apparently exceeded the capability of the detection and telemetering systems and the signal as received at the ground was interpreted as indicating operation of the detectors at counting rates above the overload level (i.e., greater than 35,000 counts per sec). The orbit of Explorer III (1958 Gamma) was similar to that of Explorer I, the internal instrumentation was also similar with the addition of a tape recorder for storage of observations for rapid repeat to the ground telemeter receivers upon interrogation. Analysis of the information obtained from these two satellites showed that the "radiation" consisted at least in part of energetic particles, presumed to be protons and electrons in geomagnetically trapped orbits.

Meteorological Measurements from an Earth Satellite

Objectives: This project undertaken at the U.S. Army Signal Research and Development Laboratory, Fort Monmouth, provided for the development of instrumentation to observe terrestrial cloud cover and other "surface" features from an earth satellite. These observations would provide preliminary assessment of how satellite experiments could be used for meteorological purposes and were expected also to be of immediate importance to synoptic meteorological forecasting.

Results: The equipment was flown on Vanguard II (1959 Alpha) with an orbit of 33° inclination, perigee of about 560 km and apogee of about 3,300 km. The good performance of the satellite sensor system indicated great detail in the variations of the reflected earth radiation received by the satellite and it was shown that these details corresponded to observations of meteorological systems scanned by the satellite. The experiment showed . . . that the simple techniques employed were capable of yielding information of sufficient resolution to be useful, and the experience gained from this flight was of high value in designing and carrying out further meteorological satellite experiments.

Interview with Rebecca Charbonneau, PhD

Science Historian, Jansky Fellow at the National Radio
Astronomy Observatory (NRAO)

Mark O'Connell: What does a science historian do?

Rebecca Charbonneau: I work at the National Radio Astronomy
Observatory, and my role is multifaceted. As a Jansky Fellow, I'm a
postdoctoral fellow. A lot of my work involves doing research, right. So
we have a historical archive in the National Radio Astronomy Observatory. And this is, it's kind of the de facto archive for all American radio
astronomy. And so I work within those archives, I conduct research
on the history of U.S. radio astronomy, as well as Soviet astronomy.
And I write books, I write papers, I give public lectures, I serve on
committees with scientific institutions, for example, I've done a lot of
research on the history of telescope siting. So the choices we make to
put telescopes in particular locations, which is a very complicated topic
and has caused a lot of contention over the years. And the idea is, is
that bringing a historical perspective can help aid decision-making in
managing and choosing telescope sites for the future.

Yeah, so NRAO is the premier American Radio Astronomy Observatory, and our mission is fairly broad. For one thing, it's to provide
world-leading telescopes, instrumentation, and data processing for
American and also global astronomers. Another part of its mission is
to train the next generation of scientists and engineers and to promote
astronomy and foster a more scientifically literate society.

Scientific diplomacy is a little bit part of my job, just because this
is related to NRAO's mission. It has a policy called Open Skies, which
it has had pretty much since day one, when it was established in 1956.
Even though we are an American observatory, we're funded by the U.S.

government through the National Science Foundation (NSF), and we're not directly a governmental entity. But definitely our mission is to make telescopes that anyone in the world can use; basically anybody in the world can apply to observe on those machines. Now, they may not get accepted, if their science questions aren't high priorities, or they're not up to a high standard, but anyone in the world, even if you don't have a degree, if you don't have an affiliation with our observatory, can apply to observe on the telescopes. And so we deal very much in a global context.

M.O.: How does IGY rate as a significant moment in the history of science?

R.C.: No historian of science, especially no twentieth-century historian of science, can ignore the fact that it was one of the most impactful moments in time for the history of science, especially for the physical sciences. But not exclusively; we also study the biomedical sciences, biology, all of these were impacted by the IGY. And so yeah, I would argue that the IGY was transformative for global science in the twentieth century.

M.O.: One of the things that appealed to me about writing about IGY was, I was just so surprised and fascinated to find out how much we didn't know, as recently as 1957. I was really shocked to discover how little we really knew.

R.C.: We lose memory, historical memory, so quickly. Actually, I talk about this all the time in my workplace. I'm always making the argument that we lose historical memory so quickly. I learned this actually, because I used to work at NASA in their history office back in 2017–2018. And this was right around the time that then President Trump signed Space Directive One, which was the great kind of the first thing that we're going back to the moon. And I remember, I remember that at the time, the NASA acting administrator was Robert Lightfoot. And he held a huge meeting and headquarters that we all attended, where he essentially said, we don't know how to go to the moon anymore. Because we don't have that institutional knowledge. A lot of the tools and equipment we used are either defunct, or the industry we worked with doesn't exist anymore. And so it's going to be really expensive. It's not a matter of just redoing something we've done. But having to almost start not from scratch but from much earlier on in the process, and it'll be very costly. And this was this kind of a testament to the value of preserving institutional memory through the historical record, archives being an example of one of the ways to do that. Because, yeah, even things that are in recent history, like the 1950s, we can lose that. If we do, we don't take active steps to preserve the history.

It's so critical. And it's also about taking active steps to preserve these documents, right? Because it's easy to just say, "Oh, well, the papers are there." But if they're not organized, if they're not processed, it makes it impossible for researchers to go back and find them. Like I have this problem doing Russian research all the time. Because often the archives are sealed, meaning people like me can't go into them and use them. So the history disappears in that sense, even though it technically exists, or maybe it doesn't, or it's there but there's no easy way to go through it. Like, I went to the Sternberg Astronomical Institute in Moscow a few years ago, and they have this attic, just crammed with papers and stuff. And they're like, yeah, probably what you need is in here, and I'm looking at this, this would be a full-time job for ten years. Like, you can't just look through it and find what you need. And I think there's not an understanding, especially among scientists who don't have to think historically, because it's not their field, they're not trained to think like this. It's not their fault. But this is serious work that has to go in, and ideally as early in the process as possible, right, to make it easier so that we don't have to hire unsuspecting graduate students to spend five years of their lives like nobody wants to do.

M.O.: Besides Sputnik, what other accomplishments did the Russians achieve in IGY?

R.C.: Well, I know they were involved in at least about half a dozen different fields of research. Nobody's looked into Soviet accomplishments in biology and medicine, and this is a very understudied part of the IGY, while also Soviet scientific history, in that I know that there was a really, really productive collaboration between Soviet and American scientists in the 1950s during the IGY regarding space medicine, because both were interested in pursuing a kind of crewed spaceflight, what was then called manned spaceflight. But also, they were very interested in kind of space biology and medicine a little bit more broadly; they did a lot more on what I guess I would call astrobiological research. In fact, the Soviets were big pioneers in astrobiology. This is part of what prompted their interest in sending up Sputnik.

M.O.: So the Soviets were looking at everything during IGY.

R.C.: The Soviets were quite interested around this period in gerontology, the study of aging, and in promoting life for as long as possible. And there were a lot of intersections between their ideas about space travel in a utopian future in space and prolonging human lifespans, and so there's a lot of weird kind of kooky research because they were also developing cybernetics around this time. The Soviets were really big on cybernetics in the 1950s and 1960s. So there's a strange intersection of all of these things.

M.O.: What is the significance of the Antarctic Treaty, and what can it teach us about collaborative science today?

R.C.: The treaty was signed in Washington, D.C., in December 1959. So a little bit after the IGY, but partly as a product of the interest that was developed in Arctic and Antarctic studies during the IGY. A big impetus for signing the treaty was the Cold War environment. And in particular, kind of an effort for arms control. There was a real concern about the militarization of Antarctic land. And the reason that I've looked into the Antarctic Treaty is largely because it was the model for the Outer Space Treaty, about a decade later in 1967. It was, in fact, based [on] the Antarctic Treaty; the Antarctic Treaty kind of served as the basis for the Outer Space Treaty and, again, had similar properties in that no one can claim sovereign land [in] Antarctica [and] into outer space. You could not militarize Antarctica as you could not militarize outer space. And what's interesting to me as a space historian is the way that in some ways, the Antarctic Treaty was more successful in being a demilitarized space, whereas outer space has always been a bit of a paradox in that, in theory, it is demilitarized. In theory, it cannot be sovereign, but you can have nuclear arms in space.

M.O.: I've been told in the course of my research that the reason the Soviets beat the United States into space with Sputnik was because President Eisenhower made it clear that he wanted to let the Soviets be first. Because he thought that would be a hugely symbolic gesture to promote the peaceful use of outer space.

R.C.: I wouldn't be too surprised. It's my opinion that the space race competition narratives have been kind of overblown. I don't think there was as much of an actual impetus to race with each other in the way that has been presented retroactively. You can see evidence of this in the historical record, also by the fact that the Soviets were just clearly more interested in doing a lot of more space activities in the earlier part of the 1960s, like we could have and just didn't. I think the competitive narrative was more of a political tool to make sure that there was public support for the funding of these projects, kind of frame it as a race. But I think the evidence is a little weak, that it was actually as intense as we think of it.

M.O.: So our classic scientific and technological rivalry with the Soviets may have been smoke and mirrors. Knowing that, I wonder whether more can be done to promote cooperative science today.

R.C.: Yeah, I spend a lot of time thinking about this. Of course, there can be, and there are, there are different ways that we do this. And all of them are fraught and complicated, because humans are annoying. So we generally think of science diplomacy—it's a term we use a lot—and people who do research on cooperative science. And there tend to be three schools of thought for the types of ways that diplomacy and science interact. The first one would be called diplomacy for science. So this is engaging in diplomatic action, specifically for the support of scientific projects. So my examples all will be radio astronomy oriented, just because that's the way my brain thinks. For example, we have a telescope called the Atacama Large Millimeter Array that NRAO helps operate with the European Southern Observatory (ESO), National Astronomical Observatory of Japan (NAOJ), and NRAO.

So these three groups all help operate it together. And they all operate the telescope in Chile. So there was a lot of diplomacy that was needed, intergovernmental lead between the collaborators, and also with the Chilean government, in order to help ALMA happen. That's diplomacy for science, right?

Then there's science for diplomacy. This is more what I write about, which is the idea that science, in some sense, like what we also call citizen diplomacy, can make an impact on a national stage or a global stage. The concept is that friendship is magic, or like the power of friendship. This is in essence what that is, in that the idea is that relationships that are formed between individuals, across national divides, can have this kind of runaway effect on the larger national temperatures, right? This is also kind of the ideal that one person can make a difference, right? We think of these as clichés, but in a scientific context, it really does work. Because scientific infrastructure and power is so important to national actors. And so we see this play out in the Cold War period, for example, in things like exchange programs, where you have, you know, a scientist from the United States visiting the USSR and vice versa, forming real relationships, collaborations with one another. And they, through their actions after the fact, either trying to establish larger-scale collaborations, earning grant money together, establishing some kind of joint ventures, they have an effect or this ripple effect of cooling national temperatures because there's an incentive to want to work with one another. In my own work, I've seen this play out a lot in the development of what's called very long baseline interferometry technique. It sounds terrible, but we had to rename it: VLC VLBI sounds boring. But you've actually heard of it before, even if you don't know the name. Do you remember the black hole image that was released a couple years ago?

M.O.: Yes.

R.C.: Yeah. So that was conducted, the reason we could do that was because of the LBI. Because essentially, the idea is that the better the res—this is kind of a simple way to put it—but in essence, the bigger telescope is either a radio telescope, [where we go by] its radius. Or, if you're using an optical telescope, the size of the lens, the bigger it is, the better the picture, right, the higher the resolution. But you can only build so big of an instrument either, not only because of money and space, but also just because of gravity, right? Like, there's only so big you can go, we're limited. But VLBI was this really creative technique that was developed by radio astronomers in the Soviet Union and the United States, and also some Canadians, who realized that if they took a telescope on one side of the planet, and one on the other, and they observed an unknown source at the exact same time, they could connect their signals and they can get a resolution that is equivalent to if they had a telescope that was the size of points between those two telescopes. So in theory, the size of half of the planet sized telescope. So it's genius. And they developed this in the 1960s. And it was kind of a great example of science, for diplomacy. Because scientists, we are motivated to have the best possible results, right?

Realize, okay, there's radio telescopes in the universe, there's radio telescopes in the Soviet Union. If we want the best results, we have to work with one another. And then they lobbied their respective national governments and were able to, you know, create the successful set of experiments, which had [a] really long-lasting effect to the point that, up until the recent events in Russian history, my observatory had relationships with the Russian government and space agencies, and that we would downlink with each other telescopes all the time. So that's a huge ripple effect that started in the 1960s, even to recent history, having these ties and connections. That's science for diplomacy. And then the last one is science *in* diplomacy. So this is probably what you're more familiar with when you think of science diplomacy. This is when we have scientists who are actually involved in diplomatic actions. Like, for example, the creation of the Antarctic Treaty, the creation of the Outer Space Treaty, things like nuclear treaties, you have nuclear physicists involved, this is the more straightforward connection between science and diplomacy. So this is the three-pronged way that scientists have a role in national relations. And I think that you need all three in order to have cooperative scientific actions.

But I also think that we underplay the personal connection, which is, I think, why the IGY was so successful, because at the end of the day, it was science for diplomacy, it was scientists working together on goals they both shared, and then using that collective power to pressure their respective governments to support their goals. So in some sense, investing at that grassroots level, just having people form relationships, form collaborations can have a much bigger ripple effect on the relations between the national powers.

CHAPTER 8

The Dawn of Space Science

To think it all started with a chocolate cake.

I have already mentioned that the idea of IGY first germinated at a dinner party in Maryland in April 1950, at which a small group of geophysical scientists had gathered to discuss the state of upper atmosphere research. It was suspected at that time that the makeup and function of Earth's upper atmosphere was a potential obstacle to manned space travel, and these six men wanted to know why.

We've already met some of the men at that party, like Lloyd Berkner and Sydney Chapman, the British astronomer who was the guest of honor. Also present that evening were Fred Singer from the Johns Hopkins University Applied Physics Lab, Wallace Joyce of the National Research Council, and E. H. Vestive of the Carnegie Institute and a former student of Chapman's. The host of this brainy soiree, James Van Allen, was the only scientist who has had an IGY discovery named after him: the Van Allen radiation belts. The hostess and chef that evening was Van Allen's wife, Abigail.

In an oral history recorded in Iowa City, Iowa, on November 18, 1997, by Brian Shoemaker of the American Polar Society, Van Allen recalled the evening's discussion in vivid detail:

> In the spring of 1950, a man by the name of Sydney Chapman, a distinguished British geophysicist, was visiting the United States and he heard about our work and he was

very interested in our rocket measurements. He came to the Applied Physics Lab one day for a visit and I went over all of our work with him. I then suggested, "Sydney, why don't we get together with some of your colleagues in the Washington area?" He said that would be great. So, my wife and I arranged a dinner party at our home in Silver Spring, Maryland. Chapman was particularly eager to meet with Lloyd Berkner and Wallace Joyce.

So, we had a very good dinner which my wife prepared and then she hustled our then two daughters off to bed while the men gathered in the small living room and shared some very good conversation. For dessert, my wife had prepared a multi-layered chocolate cake which was made of about ten thin pieces of cake-like batter, each one of which was frosted with chocolate frosting. It made a fantastic dessert which she served that evening. I thought that her dinner was one of the important bases for planning the International Geophysical Year.[1]

"It was a very pleasant and convivial evening among some, you might say, really important people," Van Allen also recalled. "During the course of the evening, Berkner spoke up after some general conversation about progress of the geophysics and various areas and additional fields of work which were being opened up these days and should be included in an international coordinated effort.

"'Well, the way things are going now, don't you think 50 years is too long to wait for the next one? Why don't we make it 25?' Berkner said."[2]

Chapman quickly observed that 1957–1958 would be the twenty-fifth anniversary of the Second Polar Year and a year of expected maximum solar activity. "That's a consideration that would make it especially interesting to many scientists," he added.[3]

And so the scientists agreed, over what we can imagine may have been very generous slices of chocolate cake, that because geophysics was "moving along so rapidly . . . the time was ripe for an international coordinated effort after a much less lapse of time than previously," Van Allen said.[4]

Looking back on Van Allen's career, it's clear that he was a bit of an overachiever. While pursuing a physics degree at the University of Iowa, he volunteered for two expeditions in the Arctic and one in the Antarctic, all aboard ship. "I have done exploratory work in both polar caps of the Earth and have been a keen advocate and follower of the scientific work, especially that in the Antarctic," Van Allen reported in his oral history. "Meteoric observations were carried on in the Antarctic as well as the magnetometer ones. So everything I did that year or two had a consequence in the scientific work in the Antarctic."[5]

After receiving his PhD in nuclear physics in 1939 from the University of Iowa, Van Allen began his career at the Carnegie Institute in Washington, D.C., where he developed what would become a lifelong interest in the study of cosmic rays. This work was cut short when the United States entered World War II in 1941, and his work was shifted to the development of the radio proximity fuze at Johns Hopkins Applied Physics Laboratory in Maryland. Sometimes referred to as the very first "smart weapon," the proximity fuze used reflecting radio beams to give anti-aircraft shells the ability to detonate when they were in close proximity to an enemy aircraft instead of needing to make a direct hit. Although Van Allen couldn't quantify the success of the devices in battle, there is no doubt that their use reduced Allied casualties while at the same time increasing the number of enemy planes brought down. "So, I did a lot of the basic research on that—the development of that fuze," Van Allen recalled. "And I was one of the co-patentees of the so-called rugged vacuum tube which is the heart of the fuze. This was long before the days of the transistors, of course. I developed one of the basic features of the vacuum tube which survived firing from a naval gun at about 20,000 g. It's a real achievement, I think, to have a vacuum tube survive that."[6]

When Van Allen completed his military service, he returned to Johns Hopkins, where he rekindled his interest in cosmic rays. This time, however, he had a bold new method of studying them that would, in time, mesh neatly into the objectives of IGY. Having learned how to pack a sensitive payload into the nose cone of a high-velocity projectile with his work on the proximity fuze, Van Allen now turned his attention to adapting the former V-2 rockets that the U.S. Army had recovered from Germany after the war to a similar function.

"It was not too long a step from building a pocket-size radar that could be fired out of a gun to designing cosmic ray detectors that could be carried into space by V-2s," wrote Walter Sullivan in *Assault on the Unknown*.[7]

"[Van Allen] pioneered the development of Geiger tube instrumentation for flight on transformed captured German V-2 rockets," reported the University of Iowa. "These flights resulted in the first measurements of cosmic rays at high altitudes above Earth's atmosphere, as well as other measurements of atmospheric ozone, solar ultra-violet light, and the geomagnetic field."[8]

Concerned that the military's supply of V-2 rockets was finite, Van Allen also was a driving force in the development of the Aerobee (a rocket similar to the V-2) and of the rockoon (a small rocket that was launched from a high-altitude balloon). Both of these technological marvels played key roles in IGY's cosmic ray surveying, work that could only be done from above the earth's atmosphere.

"I went back to the Applied Physics Lab as a civilian," Van Allen recalled. "I had earned my spurs around there, and I was authorized to build up a high altitude research group using German V-2s. I built up a group doing scientific research with high altitude rockets. Originally the V-2s and then with the American Aerobees whose development I oversaw. I was the guru of the Aerobee, yes. We started it in the autumn of '46 and had the first firing in '47."[9]

Years later, when IGY was fully underway, Van Allen served and supported the rocket program in just about every way imaginable. His exhaustive list of IGY credentials tells the story: chair, Technical Panel on Rocketry; member, Technical Panel on the Earth Satellite Program; chair, Working Group on Internal Instrumentation; chair, Technical Panel on Cosmic Rays; and chair, Technical Panel on Aurora and Airglow.

"The most important one was the one we eventually called the Satellite Panel, originally called the LPR Committee," Van Allen said, explaining that LPR was an attempt at science humor: as mentioned previously, LPR stood for "long playing rocket," a play on the commonly used acronym for vinyl music albums (long-playing records): LPs.

"Within the Satellite Panel, I was chairman of the Working Group on Internal Instrumentation (WGII). As chairman of the WGII, I was responsible for soliciting and selecting the experiments which were to be conducted on the satellite itself."[10]

According to Herbert Friedman, "The number one priority went to an instrument for which I had the responsibility. It was one which would carry ultraviolet Lyman alpha and x-ray detectors. The second one was an instrument that Jim Van Allen was instrumenting to study cosmic rays. And then down the line there were various other experiments planned."[11]

In the end, the working group selected Van Allen's own project, a "Geiger-tube" cosmic ray detector, to go up in the first U.S. orbital satellite. Remember: rockets go up and then go down, while satellites go up and stay there. To achieve its best results, Van Allen needed his instrument to stay up there, and the Soviet Union's success with Sputnik proved to the world that it could be done, even if there were delays on the American side.

The Geiger-tube device, which Van Allen referred to as "my little apparatus," was selected as the primary scientific instrument on what was to become Explorer I, the first successful American satellite, launched January 31, 1958. "The same device went up with Explorer 3 on March 26, 1958. Explorer 3 carried a tape recorder that allowed data to be obtained for an entire orbit. Data from both Explorers 1 and 3 resulted in the discovery that Earth is encircled by two donut-shaped belts of energetic charged particle radiation, now known as the Van Allen radiation belts."[12]

*   *   *

The 1961 science fiction motion picture *Voyage to the Bottom of the Sea* follows the crew of an American super-submarine sent to the Arctic to use an experimental atomic missile to destroy the Van Allen Belts, which have mysteriously "caught fire." The impending danger is reinforced by frequent shots of shimmering, iridescent clouds of atomic death hovering over the earth. While this plan to save the earth from being cooked to death may sound good when Hollywood stalwart Walter Pidgeon is explaining it, firing an atomic missile into something that is already ablaze with runaway atomic heat seems akin to putting out the proverbial fire with the proverbial gasoline. No thanks.

Fortunately, we have less to fear from the Van Allen Belts than Hollywood would like us to believe. Radiation actually tracks as less scary when it can be described as a donut, or as an ordinary fashion accessory, as represented by this exchange related by Van Allen himself: "I gave an explanation of my interpretation at a press conference. . . . I called it 'Geomagnetic radiation.' And one of the reporters stood up and was trying to visualize what I was saying, and he said, 'You mean it circles the earth like a belt?' and I said, 'Yes, that's what it is. It's like a belt around the earth.' And that's the way it got its name 'Radiation Belt' from this exchange between the reporter and myself."[13]

"Often the view of the general public about a discovery does little credit to the effort behind it. It is as if someone pressed a button accidentally and, lo and behold, there is a discovery!" wrote Doraswamy Venkatesan in his review of Van Allen's treatise *Origins of Magnetospheric Physics*. "Such a superficial perception grossly underrates the achievement, which is quite clearly a culmination of serious systematic studies that span many months or several years.

"One can feel the spirit of adventure and recognize the ingenuity that is part and parcel of scientific exploration."[14]

To say that Van Allen was at the forefront of what Venkatesan describes as "the step-by-step unfolding of the nature and complexity of the radiation belts" is an understatement.[15] By sending Geiger-tubes into space in his fleet of V-2s, Aerobees, and Rockoons, Van Allen was the first to observe cosmic rays as they penetrated the earth's upper atmosphere.

"The immense opportunity for finally being able to make scientific observations through and above the atmosphere of the earth drove us to heroic measures and into a new style of research, very different than the laboratory type in which many of us had been trained," Van Allen wrote. "In my own case familiarity with nuclear physics techniques, exposure to cosmic-ray and geophysical problems at DTM (Department of Terrestrial Magnetism at Carnegie Institute), and World War II experience in ordnance and gunnery impelled me into this new field of research."[16]

But what did Van Allen see, exactly?

Walter Sullivan described cosmic rays as the most exciting discovery of the IGY.

Cosmic rays are composed of high-energy particles that bombard the earth's upper atmosphere from all directions in space at very high speeds. And when I say "high," I mean "nearly the speed of light." And when I say "in all directions," I mean every which way, all the time, day and night. Simply put, every last star and galaxy you can see and every one you can't is constantly churning out these particles from all points and sending them off into the void of space to meet their cosmic fate. Some, but not all, of the rays that reach the earth come from our sun, which means that the vast majority come from stars and galaxies "out there," in outer space. They may be relatively few in number, but the rays originating from our sun hit our upper atmosphere at a much higher rate of speed than those from distant stars and galaxies. The vast majority of the particles that make up "our" cosmic rays never make it to the earth's surface, however, and we should be grateful for that. These particles that never reach the earth instead collide with nitrogen and oxygen nuclei in the atmosphere, creating a menagerie of particles that are still being studied today.

"The observational work that led to the discovery of the radiation belts of the earth, later recognized as a portion of the magnetospheric system, was done between 1946 and 1957 with equipment on rockets that fire more-or-less vertically to altitudes of the order of 100 kilometers. The scientific context of the discovery had a much longer and more diverse history," wrote Van Allen in his treatise *Origins of Magnetospheric Physics*. "Those of us who have participated in this work count the period that began in 1958 as one of the most rewarding in scientific history."[17]

In its special November 28, 1960, issue, *Life* magazine declared that the belts were a "magnetic trap," and "the single most surprising discovery of IGY."[18] *Life*'s feature on the earth's magnetic field and the Van Allen Belts, introduced with the headline "Two Electric Belts That Girdle Earth," was one of a series of four articles illustrating and explaining the amazing results of IGY.

On a two-page spread with bold and colorful graphics, *Life* depicted masses of charged particles from the sun, shown as pink and blue squiggles held together by a translucent cloud of energy, careening through space from the sun to the earth. The pink squiggles represent positively charged protons, and the blue squiggles are negatively charged electrons, both ejected by the sun's flares.

In the path of these squiggle clouds is the earth, hemmed in around the equator by, yes, two "doughnut-shaped rings of charged particles." The two rings, the inner and the outer, are nestled together in alignment more or less with the earth's equator. In the *Life* illustration, both belts are shown as cutaways, so the reader can see the inner and outer lines of force, the outer ring being the glaze on the inner doughnut, so to speak. A block of text explains that the "harmful radiation" emitted by the two belts could be a dangerous obstacle to future astronauts.

Above and below the rings, strong and weak cosmic rays, represented by large pink proton squiggles and blue electron squiggles, come at the earth from outer space. Only the strong rays penetrate the belt, while the weak rays fade away when they encounter the farthest reaches of the outer Van Allen Belts. Tip the whole thing on its edge, and it looks like the tire of a car, with the earth being the hub, the inner belt being the innertube, and the outer ring representing the tire itself.

When the particles encounter the earth's magnetic field, a day or two after being cast off by the sun, they swerve into the outer line of force of the outer belt and then converge near first one pole, then the other. The inner Van Allen Belt hovers twenty-five hundred miles above the equator, and its particles jump from pole to pole, piercing the magnetic field and reaching the earth's atmosphere. Weak rays are deflected, while strong rays smash into the atmosphere and create neutrons and uncharged particles. And on and on it goes. The *Life* story ends with the final stage of the process, in which protons and electrons, trapped by the earth's magnetic field, live out their lives drifting eastward and westward in the inner belt.

At the time of IGY, geophysicists concentrated on the slower cosmic ray particles, as the slower particles were most influenced by the earth's magnetic field. Also, the intensity and duration of the upper atmosphere collisions could be studied by IGY scientists at forty-some observation posts using the recently invented neutron counter.

If you were an astronaut preparing for an orbital flight, you might be worried about passing through such magnetospheric chaos, and you would be right to worry. There's a reason Van Allen once described "his" belts as being "Wild and Wooly."

According to the NASA website, "One of the largest hazards for astronauts traveling to Mars will be overcoming exposure to high energy radiation from the solar wind, solar storms, and galactic cosmic rays that originate outside of our solar system."[19]

"The Earth's magnetosphere traps the high energy radiation particles and shields the Earth from the solar storms and the constantly streaming solar wind that can damage technology as well as people living on Earth," the website continues, pointing out that the outer Van Allen Belt is made up of billions of high-energy particles emanating from the sun, while the inner belt is spawned by billons of interactions between cosmic rays and the earth's atmosphere.[20]

Despite the dangers, Van Allen theorized that astronauts and spacecraft could safely travel into outer space by flying through the weaker portions of the belts. Sure enough, in 1968 NASA's Apollo 8 moon mission was the first manned mission to pass safely through the Van Allen Belts, resulting in many sighs of relief at NASA's Mission Control and at least one in Van Allen's University of Iowa office.

"Certainly, the discovery of the radiation belts was the most important discovery of the International Geophysical Year," said NASA physicist George H. Ludwig, PhD, in 2000, "because it represented a discovery of some major phenomenon that had a substantial impact both on scientific research and on plans for manned [space] flight later on."[21] This analysis, from a scientist who actually built scientific instruments for the Explorer spacecraft that proved the existence of the belts, carries some weight. Indeed, the discovery that the earth was encircled by donuts of radiation came as a shock to many, including Van Allen's research colleague, Ernie Ray, who, after looking over Van Allen's experimental data, blurted out, "My God, space is radioactive!"[22]

* * *

More about radiation: Studying the radiation surrounding the earth led to the discovery of the famous Van Allen Belts, but the work that led up to that discovery was no less impressive, according to the records of the 1965 *Report on the U.S. Program for the International Geophysical Year*:

Summary of Program

The geomagnetic program consisted of a series of experiments mainly designed to yield facts about the magnetic field fluctuations which arise from influences external to the surface of the earth. Projects provided for the procurement, installation and operation of geomagnetic equipment in several new stations and for the operation, collection and analysis of data from new and existing geomagnetic stations.

East-West Network of Magnetic Recording Stations

Objectives: Inasmuch as there had been some interest during the past decade or so in possible correlation between geomagnetic activity and meteorological changes—possibly through mass movement of atmospheric "cells" and possible differential ionization in the high atmospheres—it was expected that the east-west chain of magnetic recorders installed by the U.S Coast and Geodetic Survey would yield magnetic data useful in such studies.

Results: Records from this project have been furnished to research scientists and significant papers were published in the scientific literature concerning magnetic storms and magnetic effects of high-altitude nuclear explosions in the Pacific (actually not part of the IGY program).

North-South Network (Alaska) of
Magnetic Recording Stations

Objectives: The original intent of the north-south array of magnetic recording stations installed by the U.S. Coast and Geodetic Survey was to supplement the three standard Alaskan magnetic observatories—Barrow, College, and Sitka—in obtaining a general coverage of variations of the geomagnetic field in and across the auroral zone. This would permit the postulation and study of electric currents in the ionosphere that would cause or be associated with the observed vector changes in the magnetic field, leading to better understanding of all the related phenomena including auroral activity, ionospheric physics, cosmic rays, and space fields in general.

Results: Considerable study of the resulting records, together with other Alaskan magnetic records, has been made by the Geophysical Institute of Alaska and by scientists of the Coast and Geodetic Survey.

Operation of Rapid-Run Auxiliary Magnetographs

Objectives: This program of the Coast and Geodetic Survey provided for recording fluctuations in the earth's magnetic field too rapid to be resolved by the time scale of the standard magnetographs inasmuch as much research by many workers has shown that a great deal of important information can be obtained from the higher frequency components of the magnetic fluctuations.

Results: The rapid-run magnetograms are contributing to numerous studies and investigation projects, as evidenced by the requests for copies of the grams from World Data Center A. The rapid-run equipment has been in continuous operation at the permanent observatories since the close of the IGY.

Magnetic Observations at Jarvis, Palmyra and Fanning
Objectives: Temporary magnetic observatories were established under this project, and operated for about one year during the IGY, by the Scripps Institution of Oceanography, University of California, La Jolla, California, on Jarvis, Palmyra and Fanning Islands, and for one month on Christmas Island, for the purpose of studying the anomalous behavior of the magnetic field in the vicinity of the magnetic equator.

Operations: It was initially planned to make measurements over a period of about 15 months. However, it turned out that Jarvis and Fanning operated for a little over 16 months, with about 80% complete records, and Palmyra for 12 months, with 50% complete records. Most of the gaps were due to instrumental troubles, some of which required spare parts to be obtained from the mainland. The Jarvis station was closed down for three weeks during October 1958 because of the death of the observer. The Palmyra station was transferred to Christmas Island for the month of December 1958.

Invaluable assistance was obtained from the British authorities, who left intact on Jarvis the installations provided for a scientific party working in the area prior to the beginning of the project and who airlifted both personnel and supplies between Hawaii and Christmas Island. The British company, Cable and Wireless, generously made available to the project the supply, workshop and social facilities of their cable station on Fanning Island.

Data from Jarvis have been processed and published as Coast and Geodetic Survey Publication HV-JI58. Data from Fanning will be dealt with similarly. All other data from the IGY period have been submitted to the data centers and are available on microfilm.

Interview with Katsia Paulavets, MS

Senior Science Officer, International Science Council

Mark O'Connell: When I started researching my book about IGY, one of the first things I came across was the International Science Council. And I started reading things like your *Flipping the Science Model* report, and the *Mission Science for Sustainability* report, and the *2030 Agenda*. Do you even know where to start to describe how exciting it is for you and your organization?

Katsia Paulavets: Yeah, it's so exciting that that keeps me awake each night. And it's, you know, like, it's a long journey. This is a journey where many people have contributed. But I'm also glad to say that to some extent, we are coming to the point when we want to finally test the model.

Of course, mission science, it's not a novel approach. This approach has been applied in the past to other challenges, but we never applied this approach to sustainability science. So there, we are certainly excited. And more importantly, we also think it's really something meaningful, and hopefully something that really will make an impact. So science until now, it didn't have a chance really to show what it is able to do within the sustainability domain. We hope that with this process, we will be able to enable science, the best science, sustainability science, to contribute meaningfully around the globe on dealing with mobility challenges.

M.O.: It's very inspiring. How did the journey begin?

K.P.: Oh, so it's been a few years ago now. We had this idea that we do want to have global discussions between key players that support science for sustainability. And see how they could overcome the existing institutional differences in our different priorities, different practices, and see how they could work together. So in 2019 we organized the first Global Forum of funders, and we brought national research funding agencies, some foundations, development aid agencies to explore basically how they could work together to support this type of science. And basically, this was the turning point, I would say. So during that discussion, certainly, the appetite for collaboration was growing across different funding agencies. Certainly there was a recognition that there are differences. But it does make sense to overcome those differences and come together on the science.

Okay. So they basically invited us to identify areas in which science funding needs to focus. And beyond, it took a massive review exercise, we mobilized the global scientific community to identify priorities, and this is how we came up with the mission that basically called for science funders to create at least five science missions in five key areas, including climate.

And that's where we set up this global commission and technical advisory group to develop a model for implementing science missions for sustainability. And now we're finally about to launch the call for identifying those pilot science missions. Of course, the context has changed significantly as a result of military conflict now. So for the environment, the context is not very favorable. Many countries are looking very much inward, not outward.

So by the same time, it does highlight that if we are to resolve the military conflicts, we do need to solve the sustainability challenges as well. Because if you look very carefully, often the sustainability challenges are one of the key reasons for those military conflicts as well. So there is, again, the need for stronger collaboration on sustainability challenges among those different funders. So now we want to build this coalition of visionary founders who will support this. And so the journey is still there. And this year, this is our goal: to undertake this journey.

M.O.: How did the sustainable development goals fit in?

K.P.: We do have seventeen goals, as you may know. But what we do with that process, we try to understand the areas that need to become resilient, that would help to advance the implementation of all seventeen sustainable development goals (SDGs). What we came up with was that if we fix, or if we make cities sustainable, and more resilient, if we resolve the climate issue, then, basically, all SDGs will be progressing. So we identified five [goals] making progress significantly on other SDGs. And we came to the conclusion that, let's see what different regions have to say where they want to focus their efforts. The approach areas, and basically, by making those areas sustainable and resilient, there is a high probability [that] for now [it] is not about identifying scientific priorities, but it's more about how we do science, how we make sure that science is not done in a vacuum, that it strongly collaborates with key relevant stakeholders, and that it brings all key actors together toward one goal. And yeah, so it's kind of the Marshall Plan, but for sustainability now.

M.O.: Okay, you mentioned just now that there are too many scientists working in a vacuum. And in your, in your papers, you mentioned several times that there are many sciences too much isolated into silos. How deep a problem is that? Do you sense any desire by the scientific community to break out of those silos to be more open with their work, to be more collaborative? How difficult is that to achieve?

K.P.: It's a diverse kind of change. It's not just about scientists, it's also about the reward structure that we have. It's also about the incentives that our universities, our research funding agencies have, and also the way science organizes.

For the moment, you know, it's about scientific excellence. And a lot of scientific excellence is organized in different pillars, so there is a lot of excellence, but within one, there is a lot of fragmentation, basically, across these different disciplines. We are not saying that excellence in science needs to be undermined; it's not that. It's more, we need to create a mechanism where those best of each disciplines can come together and work together on a common goal on a common problem.

The science mission model is trying to do that, to provide this platform where the best of global science coming from different disciplines—and actually, not only natural science; a critical role will be given to social sciences, because with all these sustainability challenges we have technical solutions, but what we don't have is a stronger understanding of societal processes and how we can break through those barriers, societal barriers, political barriers, vested interests, and so on. So input from social sciences to science missions will be critical. And most of the science funding goes to the natural sciences, not social sciences. So again, there are structural systemic problems within science systems.

With the science mission approach, we will not solve it. We do claim that it's important to the evolution of science, but it's a longer-term process. But we try to highlight where these changes need to happen. What we argue with the science mission model is that we cannot wait any more until science systems are fixed. Because basically, with the current situation, when it comes to the planetary boundaries, we cannot wait for another ten years; we need to act now. We need to bring science and action together and act. So the science missions, it's not about new knowledge generation; it's more about how you put the best available science [in]to action, from different disciplines [in]to action. That's very exciting.

M.O.: Can science today come together the way it did for IGY in 1957? Could today's scientists duplicate that effort? Now, is that wishful thinking? Or do you think that's something that could happen, at least to some degree?

K.P.: I mean, this is exactly the point of this: science missions. But it's more about why the scientific community was looking outside of the earth. And now what we're trying to say is that there is no better place than planet Earth before going outside of planet Earth. Let's use the best of science to make this Earth even better to protect it and to ensure our human and planetary well-being.

The idea is to mobilize the large community of scientists, but of course, you know, the funding is limited and what needs to be understood is what was behind IGY. There was tension between East and West, and there was strong economic and military interest behind them. So there were many spin-offs also of that project, you know, the Antarctic Treaty was created, and so on. So there are many, many things that have a long legacy. So it is for us to create something ambitious, that would also have that long-term legacy. So the ambition is there, but then, you know, whether the funding will join, this is another thing.

At the same time, the context is similar to some extent. But the challenge is that there was strong economic and military interest in the IGY endeavor. Often the value of doing or acting for the global common good is not obvious to many, or, unfortunately, you still need to put the economic value on that. Of course, it's not straightforward for many people, or it's straightforward for decision-makers, but they actually use [calculations of value] as an argument not to do it.

M.O.: How do we get past those obstacles that you're talking about? And you're mentioning funding a lot. That's probably the number one issue, but there are other issues as well. How do we get past the obstacles as a society?

K.P.: This is the key question that everyone is thinking. I still, despite fifteen years of experience working in sustainability, I do believe in societal commitment, and people basically, there are good people. We just need to build that coalition of willing partners, and they are there. There are many people willing to join and come together and, and more importantly, I think, also experiment. Because to me, action is critical, but we often stop because we think too much. And we act little, you know, we overthink sometimes. And, and I think if already we could put the knowledge that we have into action and experiment and see whether it would work or not, it gives inspiring stories for others to join, to replicate, to do differently. But at least to do and to try and yeah, I know that any initiative starts at the small scale. And then if it's successful, it will grow, and if it not, then we will learn what we can do better and try again. The key is to experiment and not to give up on our belief in humanity.

CHAPTER 9

The Sky Above

One of the most interesting and entertaining aspects of writing a book like this is the way the topic of the book starts to resonate as you seek out people to meet with and to interview. I'll be honest; even though my publisher had confidence in me and my idea, I did not know at first whether the idea for this book would fly at all. Would the scientists and scholars I approached find the idea interesting and of some value, or would they find it naïve, simplistic, even laughable?

I'm happy to say that everyone I've interacted with felt that the idea had merit, even if they didn't agree with some of the assumptions I had made from the outset. Luckily for me, I kept finding scientists and scholars who are able to find value in discussing a topic even if they find fault with it. How did I meet these supportive scientists and scholars? Well, in a sense, they found each other for me.

At several crucial moments in the writing of this book I would be winding down a fascinating interview only to have the interviewee say, "You know who you should talk to is . . ." followed by the name and impressive credentials of someone I *really should talk to* about IGY and climate science. When someone made a recommendation, it sent the message to me that the person I had just interviewed must have felt that I was a decent interviewer, knew the best questions to ask, and was well worth talking to. That made for a nice ego boost. A couple of the people who were recommended to me had passed on, one or two weren't interested in being interviewed (which is always disappointing but also always respected), but most were more than happy to talk to me.

One recommendation in particular nearly escaped my attention, as it came up at the tail end of my interview with Amanda Townley, executive director of the National Center for Science Education. Somehow it didn't register with me until later, when I was transcribing the interview with Director Townley, that she had mentioned someone I should contact: "Do you know Marshall Shepherd?" she asked. "He might be a good person to talk to as well. He's one of our board members. He's also a Distinguished Chair in Atmospheric Sciences at the University of Georgia. But he has worked extensively with NASA, and he was tapped by NOAA to take on a leadership position quite a few years ago. He is a phenomenal person and just a wealth of information. And if he's not the person, he can definitely point you to someone who is!"

Guess what? He was the person. Still is, in fact.

When Dr. Shepherd was able to squeeze in an interview between meetings, the talk quickly turned to IGY and its influence on his area of specialty, the science of meteorology. Dr. Shepherd spent twelve years at NASA, working extensively on the types of satellites that point down, and so was keenly aware of the role that IGY played in the development of the new science of "space weather."

"IGY certainly is sort of a rallying event for understanding our planet and its complexity," he told me. "For sure, it is seen as a guiding principle. I do understand and appreciate its importance in the broader, big picture sense of guiding thought and organizing principles to galvanize policymakers, decision makers, stakeholders and the public to think carefully about this planet for which we don't have a Plan B."[1]

"If you think about the time that IGY happened, that was also around the time that NASA was founded," Dr. Shepherd went on. "NASA was formed in 1958 and a lot of people thought of NASA as just the moon and space shuttles and things. But NASA has, and always has, and even today has a very strong focus on using its capabilities to look back at the third planet from the sun, Earth."

Dr. Shepherd explained to me that space weather couples the behavior of the sun, CMEs, ionized particles that come from the sun, and the earth's atmosphere, including our ionosphere and magnetosphere. "And so, there's a really interesting coupling of those two environments," he said. "As you know, we had a brilliant example of that recently, sitting here in my front yard looking at the aurora borealis here in the metro Atlanta area, which is just unheard of."

The unheard-of occurrence of the aurora borealis in metro Atlanta was due to a significant CME event associated with unusual sunspot activity—a perfect example of a dynamic that is foundational to the important subdiscipline of space weather. "Scientists who study space weather are trying to understand the impacts on solar activity and solar phases, and how men on earth affect atmosphere processes, and now geomagnetic processes, and how that affects society," he said, hitting home our dependence on GPS and the electrical grid and explaining how those types of significant space weather activities can "very much disrupt" those anthropogenic or human activities.

Dr. Shepherd's career as a scientist was launched by a bee sting. An only child, he spent much of his youth in his backyard catching insects, which led to a keen interest in science and nature. But one day he got stung by one of those insects and discovered that he was highly allergic to bee stings.

It didn't take him long to abandon his insects and find a new avenue for his scientific predilections. "I switched my sixth-grade project, and I did a science project called, 'Can a sixth grader predict the weather?' And I won the science fair." Weather became his new passion, and he began building simple hygrometers (used to measure the moisture content of the air) and balloon barometers out of household materials.

"And so I was bitten—pun intended, I guess—by the weather bug at that point. But I never wanted to be a forecaster predicting. I was always more interested in the hows and whys of our weather and climate."

Dr. Shepherd went on to spend a dozen years at NASA as a research meteorologist working on large weather climate satellite missions to study the earth's weather and climate and how it's coupled to oceans and the biosphere and cryosphere, "and all of the spheres," he quipped. Ultimately, he joined the faculty at the University of Georgia as the director of the atmospheric sciences program, where he has been for eighteen years. Along the way he served a term as president of the American Meteorological Society (AMS), the largest professional society in atmospheric sciences.

"The program I'm the director of here at the University of Georgia's atmospheric sciences program includes meteorology, climate sciences, atmospheric pollution, atmospheric optics, and so forth," Dr. Shepherd told me. "We're interested in the predictability and understandability and challenges from the sort of time scales of weather out to the long-term climate time scales of climate change and everything in between, including things like sub-seasonal and inter-annual variability, things like El Niño."

If Dr. Shepherd's work sounds to you like the IGY meteorology program writ large, you're not alone. Throughout our conversation I was struck by the realization that Dr. Shepherd is wrestling with some of the very same issues that occupied climate scientists in 1957, only dealing with way more data and way more human influence. It certainly seems that if not for IGY, Dr. Shepherd would have a vastly different career today.

*   *   *

These days the National Oceanographic and Atmospheric Administration (NOAA) monitors the air above and the water below, which brings up a question I have pondered throughout the process of writing this book: Why do we have one governmental agency to study two fields of science, and why these two? I mean, it makes sense for NASA;

aeronautics and space are so closely related, and with so much overlap, both working to counteract the force of gravity so that we can traverse and explore the outer reaches of the atmosphere and beyond. Health and Human Services seems the same, and I'm sure there are more that I'm not thinking of. But oceans and the atmosphere seem pretty fundamentally separate to this nonscientist.

So I sought out the opinion of a scientist who is well qualified to answer my question—marine biologist, author, and expert on sea level rise, Dr. Wallace J. Nichols. Why, I asked him, is one government agency responsible for both oceans and atmosphere? What's the connection that I'm failing to see?

"They are essentially the same thing," he said of the ocean and the atmosphere. "Air is water. It's a continuous entity, really."[2]

Somewhat bemused, Dr. Nichols noted my fascination with the concept of a globe-encircling ocean and air continuum, so he ventured deeper into what for me were uncharted waters: "It *seems* that ocean and atmosphere are strange bedfellows, but clouds are, in fact, water. Air is, in fact, water. People don't recognize it as such, but I can look at my phone right now and tell you the humidity outside," he said. "That's the water content of the air."

Dr. Nichols went on to explain that he has flight trainer friends who describe flying to their student pilots as coursing through water, moved by propellers that function like a paddle pushing a boat forward. "You can understand the physics of flight much better this way than if you have this confusing idea that the sky is nothing. It's not. It's far from nothing; it's a viscous environment."

"I would echo that in some sense," Dr. Shepherd told me in our later conversation. "The oceans and the atmosphere are both fluids and they're governed by the same equations. There's this continuum. The atmosphere is a fluid, and it's on a rotating body. And so this fluid, we look up and take it for granted, but the atmospheric fluid and the weather processes and climate processes associated with it dictate almost every facet of our day, from agriculture to energy production to disruptions of our soccer games.

"When people study atmospheric sciences and meteorology or oceanography, I think there's this sort of whimsical interest in clouds or storms or ocean currents or so forth. But, the atmosphere and the ocean are governed by fluid mechanics, fluid dynamics and thermodynamics. So, you've got these two fluids on a rotating body, the earth, and if you understand one system, you understand the other system in many ways, because not only are they similar or the same, they're very much interconnected.

"People often are stunned when I say that. You know, if we're in an El Niño or a La Niña, that affects the hurricane season in the Atlantic Ocean, but El Niño and La Niña are patterns that happen in the Pacific Ocean," he told me. "El Niño is when the waters of the eastern Pacific are warmer than normal. La Niña is when the waters are colder than normal, and that affects the air column over the water, which in turn affects jet stream patterns and the wave patterns in the atmosphere that ultimately affect our weather and whether we're going to have more hurricanes in a given season."

Dr. Shepherd teaches his climate students that the earth is always trying to put itself on what he refers to as a "heat diet." The earth, he tells them, has too much heat around its middle, much as you or I may have unwanted weight around our midsections. Because we have this disproportionate amount of heat in the tropics, Dr. Shepherd said, the earth is always trying to redistribute that heat to the poles, and both the oceanic and the atmospheric currents help to make that happen. "So, if you look poleward, flowing currents or warm currents, things like the Gulf Stream off the eastern coast of the U.S., it's warm water. Hurricanes are moving moisture and heat from the tropics to the poles as well, and many of our sort of atmospheric weather circulation patterns, so they are working together as a part of this large heat diet."

As observed in *Life* magazine in 1960, "Heating the atmosphere from below rather than—as many people believe—from above, that helps power the weather."[3]

"Weather prediction and short-term climate prediction have always been there and sort of had some of its glory days during the time of IGY," Dr. Shepherd told me, adding that some of the first applications of the world's very first computer were to "crunch" weather forecasts because the equations were so complex.

Dr. Shepherd added that the weather and climate processes that he studies are fundamentally tied to national security, the economy, energy, water supply, and more. "So, whereas the sixth grader and maybe even the college student in me was just curious about, why does that hurricane form, or why does it get so much stronger? Why does that storm produce a tornado? . . . [T]he more seasoned scholar in me understands that these very processes that I was so curious about as a child, very much affect every aspect of our lives." He pointed out that increasingly over time, there are disproportionately more people at greater risk from a hurricane that makes landfall in Houston (as we just saw in the summer of 2024) or from the 120-degree heat that has blanketed the western United States.

*   *   *

Many of us have at one point in our primary education learned about the water cycle: condensation, precipitation, evaporation, and then back to condensation to start the cycle all over again. All pretty simple and understandable, right? Well, Dr. Shepherd had a surprise for me: the traditional water cycle, he claimed, is flawed.

Here is his explanation. "There's been a widening of our scope to understand the broader challenges that the atmosphere and ocean-coupled system conveys, but also how those two systems interact with other parts of the Earth system, whether it's the biosphere or the human-coupled natural system. Humans," he said, "are increasingly a part of the atmospheric sciences narrative."

But, strangely, the traditional water cycle does not take human processes into account. "None of us has ever lived in a water cycle that behaves the way we learned it," he said, "because we've always had reservoirs and pollution and irrigation and piped in water into cities and other sorts of human activities that we fail to consider" (more on this topic later).

When I asked Dr. Shepherd whether humans harbor any other misconceptions of how the atmosphere works, he had a big one: "Weather and climate aren't the same," he declared. "I often say, 'weather is your mood and climate is your personality.' And I say that because people often get confused about weather. So, on a cold, snowy day in January in Vermont, some guy on Twitter will tweet me and say, 'It's snowing today. What do you guys mean? Global warming and climate change?' That's a fundamental indication that person doesn't understand the difference between weather and climate. A day's worth of weather, or even a week's worth of weather doesn't define climate in any way. So, I think that's a misconception.

"I also think there's a misconception that humans cannot affect the climate," he concluded. "I have people all the time tell me that the climate always changes, it's naturally changing all the time. And I say, 'Absolutely it is. But, grass always grows naturally, too. And when we fertilize our lawns, it grows differently.' So, I try to remove this narrative that it's an either-or proposition. There are naturally varying processes in the atmosphere and ocean, and human activity can affect them."

*   *   *

It will probably come as no surprise to the reader, as you've seen from my interactions with Drs. Nichols and Shepherd, that in the course of researching IGY over a year and a half, I encountered a great many scientific surprises from a great many directions, findings that challenged my view of the world and of outer space, some so stunning that I still find them hard to believe. I had never considered, for example, that the sun had an atmosphere of its own so big that it envelops the earth, or that there were auroras in both the northern and the southern skies that appeared and behaved in lockstep with each other, or that the oceans and the atmosphere were virtually the same thing, or that our continent was constantly drifting around the earth's surface at about the same speed at which our fingernails and toenails grow.

I mean, that actually seems *speedy* to me for a whole continent.

The biggest surprise came when I read "New Portrait of Our Planet: II," the second of four installments in *Life* capturing in words, photographs, and paintings the amazing scientific adventure that was IGY. Published in four consecutive "chapters" throughout November 1960, "New Portrait of Our Planet" presented readers with spectacular images of Earth's drama and grandeur that still stretch the imagination. I've referenced the *Life* words and illustrations elsewhere in this book, but I think they have earned greater, more focused attention here. In one photograph from chapter II, for example, an IGY scientist hangs on for dear life on the deck of the oceanography vessel *Spencer F. Baird* as a violent wave crashes across its deck. In another, more subdued photograph from chapter IV, a monstrous metal mesh radio telescope—described by *Life* as "bizarre"—listens to the silent sounds of charged particles hurtling into the earth's atmosphere from a solar flare. In my favorite illustration, conceived by artist Antonio Petruccelli in the first part, a massive underwater avalanche is depicted as crashing violently to the ocean floor and spreads out at a speed of 60 knots, endangering the nearby transoceanic cable, in which a hapless and perhaps doomed whale has entangled itself. It's a lot of action in one image, but it deftly illustrates the vast global scope of IGY.

The big surprise I'm referencing now appears in the first spread in chapter II, subtitled "Surprises of Seas and Skies: IGY Discovers Huge Streams in the Air and Ocean Deeps." After a concise reintroduction (remember, this is two years after IGY) and a brief overview of the size and scope of the event, *Life* concentrates on the many scientific revelations in the study of the earth's atmosphere that have only come to light because we have been able to put artificial satellites into orbit due to IGY.

"Compiling their findings, IGY teams produced the first consecutive daily weather maps of the globe," *Life* related. "They observed that most glaciers are shrinking and the Arctic is melting. They discovered that Antarctica contains 40% more ice than they thought, but that this ice-locked land is slowly warming. Over-all, International Geophysical Year studies of wind and water, of the earth's heat and its ice, confirm the fact that the world's climate is getting warmer."[4]

Let's read that again: "The world's climate is getting warmer." In November 1960. That's an amazing date to me, especially since I was born in May 1960. We already knew around the time I was born that the earth was on a trajectory to heat up. Simply amazing.

Then further IGY research along the same vein confirmed *Life's* diagnosis: "There is a stimulating prospect ahead of us: that mankind is quite unwittingly altering the overall climate of the earth, by spewing carbon dioxide into the atmosphere in ever-increasing quantities, from factory chimneys, industrial plants, and car exhausts," wrote Ronald Fraser in his 1958 book *Once Round the Sun*. "Glaciers in the northern hemisphere are shrinking. In Scandinavia they are drawing back up the mountain valleys, in Greenland their ocean terminals break south in sea-borne ice, in Switzerland there is worry about future sources of water-power. Is this trend world-wide, evidence of a slow warming of the whole earth? Here Antarctica can perhaps answer at any rate the last of these pressing questions. For around 90 per cent of the total ice-cover of the world lies on the southern continent—the glaciers of the rest of the earth are an insignificant 10 per cent. Thus a determination of the direction of the trend of the expected shrinkage in the antarctic ice—which may of course turn out to be a confirmation of its stability—is of more than merely academic interest to mankind; for wastage of all the enormous ice-sheet of Antarctica would raise the average sea level by nearly 200 feet. This might conceivably happen in say 10,000 years; but an initial rise in the average sea-level of say 30 feet in 10 years is not impossible."[5]

Ten years. 1968. A 30-foot sea-level rise was "not impossible." And what were we doing about it?

Journalist Walter Sullivan also took note of the warming climate in his book *Assault on the Unknown*, stating that "the Swiss Glacier Commission reported, for its 1957–58 observation period, that eighty-three out of eighty-nine glaciers had retreated an average distance of forty-eight feet."[6]

"The Arctic sea apparently has been growing warmer," wrote Alexander Marschak in his 1958 IGY book *The World in Space*. This portends ill winds indeed "if, as the meteorologist thinks, the earth is warming."[7]

Two years later, *Life* magazine gave plenty of reason to be concerned about not just what the meteorologist thought but also what the meteorologist now knew for a fact: "The temperature of the Pacific has risen two degrees F within the last three years. The economical effect of this change has been substantial. It has bankrupted the guano industry of Peru and Chile, which sells fertilizer made from coastal bird droppings. Plankton, which thrives in cold water, is no longer abundant off the coast, so plankton-eating anchovies have moved away. The birds which eat the anchovies have starved.

"Because of the warming water the California salmon industry has been ruined and that of Canada seriously afflicted when many of the cold-loving salmon died."[8]

That's a shocking amount of global warming awareness—and damage—already by 1960, but the awareness goes back even further, to the days just after the end of World War II, and it involves the military. "Early concern over a subfield of global climate change—polar warming—came not in the 1960s or 1970s, as the U.S. environmental movement blossomed, but in the late 1940s," wrote geophysicist Ronald E. Doel in a 2002 article. "The issue was discussed at a secret Pentagon briefing where evidence from the Swedish climatologist Hans Ahlmann that glaciers were receding in high latitudes was scrutinized."[9]

According to Ronald Doel, the U.S. military's interest in the thickness of the polar ice cap was driven by two factors: first, the Arctic ice sheet gave the Soviet Union an all too easy approach to the northernmost U.S. defenses, and second, the U.S. Navy's Polaris class submarines were designed to break through the ice sheet to launch their nuclear missiles, so the thickness of that ice was a crucial issue in the effectiveness of the submarines. "Through the first half of the Cold War, global warming was a strategic concern for Pentagon officials," Doel wrote.[10]

The scientists of IGY had another reason to be keeping a close eye on the ice caps, but it had more to do with reflectivity than with thickness. One of the more fascinating (to me) portions of the IGY meteorology program involved observation and monitoring of the earth's *albedo*, a term that describes the amount of sunlight reflected off an object, like, say, the earth. In other words, IGY was studying "earthshine."

The sunlight that penetrates the barriers of our atmosphere and reaches the earth's surface will do one of two things: if it shines on a dark surface, such as thick forest, most of its energy will be absorbed by that surface. The average albedo for a thickly forested surface is about 15 percent, meaning 15 percent of the sunlight that hits the forest will be reflected back into space, and 85 percent will be absorbed. If sunlight shines on a bright surface, such as a glacier or an ice cap, 90 percent of its energy will be reflected back into space, giving snow and ice an albedo of 90 percent. Due to the many textures and colorations of the earth's surface—including clouds, land, water, and ice, each with its own specific albedo—Earth's planetary albedo averages out to a comfortable 31 percent, indicating that nearly a third of the solar energy that researches our planet is sent right back into space.

What this means in terms of climate change should be easy to grasp. If the earth's albedo rises or falls in significant measure, there may be serious consequences, according to the University Corporation for Atmospheric Research (UCAR) Center for Science Education, which warns, "If Earth's climate is colder and there is more snow and ice on the planet, albedo increases, more sunlight is reflected out to space, and the climate gets even cooler. But, when warming causes snow and ice to melt, darker colored surfaces are exposed, albedo decreases, less solar energy is reflected out to space, and the planet warms even more. This is known as the *ice-albedo feedback*."[11]

"As glaciers, the polar caps, and icebergs melt, less sunlight gets reflected into space," NASA explains at its Scientific Visualization Studio. "Instead, the oceans and land absorb the light, thus raising the overall temperature and adding energy to a vicious circle."[12]

Project 8.11, Earth Albedo Observations, was organized by the Smithsonian Astrophysical Observatory "in order to obtain information important to the study of the earth's atmosphere," as described in the *Report on the U.S. Program for the International Geophysical Year*, published in 1965.[13] It is pointed out in the report that the albedo or reflectivity of a planet can reveal much about that planet's atmosphere. For example, the planet Mercury has an albedo of 0.06, while Venus has an albedo of 0.64, indicating that the two planets closest to the sun have dramatically different atmospheres.

"The albedo of the earth cannot be deduced from observations analogous to those that give the reflective power of the other planets, because we cannot see the earth as a whole," the report continues. "Indirect information, however, is given by the earthshine on the moon, such measurements are difficult because they can be made only near new moon at which time the moon can be seen only at low altitude. Large corrections for atmospheric absorption and twilight illumination have then to be taken into account."[14]

This difficulty was surmounted in large measure by the use of a Danjon Prismatic Astrolabe, a device that had recently been invented by French astronomer André-Louis Danjon, "one of the leading French Scientists of the IGY."[15] An astrolabe is sort of an astronomical Swiss Army knife, used by astronomers for centuries to do everything from measuring the altitude of a celestial body to charting the lunar calendar. That Danjon was able to significantly improve an instrument that had been in common use since 150 BC is nothing short of stupendous. That the improvement was driven by the necessities of IGY was just icing on the cake.

Project 8.11 was organized and carried out under the watchful eyes of astronomers G. A. Van Biesbroeck and J. Allen Hynek, who would soon become a nationally known figure for being the first U.S. scientist to get the news that Sputnik had been launched into orbit. The objective of Project 8.11 was to "carry out measurement of the earth's albedo during the IGY from a network of six earth satellite optical tracking stations using Danjon's measuring tool. The albedo is found from the brightness of the earthshine on the dark part of the moon's crescent disc. An instrument compared this brightness with that of the crescent adjacent to it, properly reduced in intensity, thereby eliminating the need of taking into account atmospheric absorption.

"Danjon . . . has found indications of seasonal changes in the albedo which he attributed to fluctuations in meteorological conditions. To investigate this question further, it was necessary to take measurements from several stations widely distributed in longitude and latitude around the world.

"The IGY offered an excellent opportunity to carry out a comprehensive earth albedo program. The general atmospheric conditions were better known than ever before. The satellite optical observation stations would give appropriate distribution in geographical position and, since they were in operation for the duration of the IGY satellite program, a sufficient amount of data could be obtained. Instruments were installed at: Organ, New Mexico; Williams Bay, Wisconsin, Mt. Haleakala, Hawaii, Arequipa, Peru, Shiraz, Iran, Woomera, Australia."[16]

Preliminary results from Project 8.11 presented at an American Astronomical Society conference in Mexico City showed that seasonal changes of albedo occur "as a result of prevailing weather conditions over extended regions," the IGY report reveals. "However, daily variations of albedo are far more important and reflect the instantaneous weather picture on the earth. The final results, essentially, confirm the preceding statements. . . . It is shown that the observations, particularly the fluctuation, have to be interpreted carefully and account must be taken of the general cloud cover and the motion of the earth's hemisphere as seen from the moon in order to relate the measured albedo and its fluctuations to changes in cloud cover. The analysis suggests that the brightness of earthshine is most sensitive to fluctuations in the range between 30 to 50 percent of cloud cover. It has been shown that the present network of meteorological stations is inadequate for obtaining the global weather picture since the cloud cover situation over the earth's water surface remains unknown. It has been suggested that earthshine observations be repeated now that a number of weather satellites can plot the cloud cover adequately."[17]

The far-ranging albedo study was but one element of IGY's thirteen weather studies. A brief review of the weather studies of IGY illustrates not only the depth and breadth of these activities but also the vast international scope of the proceedings:

- Ongoing surface observations focusing on geochemistry and radiation were to be undertaken at Arctic Ice Island T-3. Existing weather stations in the eastern United States would form a chain of observer posts to the Caribbean, while a hurricane network continued the chain of stations to South America via Balboa, Canal Zone. A significant gap in the chain was in South America, where the entire east coast was devoid of observation stations. Ultimately five cooperative sites were established: three in Chile and one each in Peru and Ecuador.
- To handle and coordinate the massive amounts of data to be generated, Antarctic Weather Central was established as a site for central data collection and analysis. The station was to be built at Little America and would be staffed by U.S. meteorologists, as well as guest workers from other cooperating countries.
- The Weather Bureau would conduct full upper air observations at all stations, focusing on geochemical and radiation measurements.
- Three world maps were to be prepared for IGY: two covering the hemispheres and one for the equatorial regions. The United States would be responsible for the Northern Hemisphere map, while the Federal Republic of Germany took responsibility for the equatorial section and the Union of South Africa would oversee the map of the Southern Hemisphere.
- Meteorological reconnaissance flights would be conducted over routes that were believed to be operational, feasible, and scientifically desirable.

- It was noted that in the program of the Union of South Africa, plans were started for twenty-four flights southwest from the Cape, while the United States proposed adding a track south-southwest from southern California. This program would include meteorological reports from commercial transport aircraft on similar flight paths.
- The Weather Bureau proposed methods to organize and collate certain meteorological data for the use of researchers. This data would be domiciled in the World Data Center for Meteorology, making it more useful to researchers. The Data Center would be located at the National Weather Records Center at Asheville, North Carolina. Data was to be recorded onto microcards, and two sets of data would be archived, one for retention at Asheville and the other to be available on loan by WDCA to institutions outside the Washington area.
- The Weather Bureau would conduct ozone measurements involving automatic chemical titration. The program would include regular air sampling for the measurement of carbon dioxide.

That last point is key: In the mid-1950s, it was generally agreed that rising carbon dioxide in the atmosphere was of scientific interest, and that reliable measurements of CO_2 were of some value, but according to astrophysicist Spencer Weart, "Nobody thought the problem was particularly important."[18]

"Early studies of atmospheric CO_2 were strictly a matter of satisfying general scientific curiosity," Weart wrote. "Their funding came from the usual sources for university research. Individuals would work on CO_2 for a few months, supported on their salary as a professor, with perhaps a little help from a government grant awarded mainly for other matters."[19]

We all know, as the IGY scientists knew, how the weather works when it's working the way it's supposed to (or at least the way we imagine it should work, the way we *wish* it would work). Our atmosphere is made up mostly of nitrogen and oxygen, along with carbon dioxide and water vapor, and a few other gases in smaller amounts.

The previously described water cycle begins—or ends, depending on your perspective—with the air absorbing moisture from the oceans and carrying it, often great distances, over land and sea. It continues with that same moisture dropping back to earth as some form of precipitation. It's a perfectly simple cycle, but, as we know, it's also a cycle that often thrives on extremes and can cause great peril.

Here's an interesting way to look at the water cycle. J. Tuzo Wilson wrote in his book *IGY: The Year of the New Moons* that the circulation of moisture between the earth and the atmosphere is what makes our planet habitable. "Without the rain which it transports, the land would be a desert," Wilson wrote. "Only fish would have developed, and the extremes between the equator and the poles would be even greater than they are."[20]

That's sobering talk, followed up by yet another revelation from IGY observer Wilson, who was keenly aware that the carbon dioxide in our atmosphere might "be effecting a change in our climate." At the time Wilson was issuing this soft warning, in 1961, there were roughly 2.4 million tons of carbon dioxide in the atmosphere, and that was up by at least 10 percent over the amount in 1900.

Wilson speculated that the burning of fossil fuels might *not* have been the reason the earth's atmosphere was warming, or might not have been the only reason, but he also referenced a colleague who theorized that carbon dioxide in the atmosphere was increasing because of deforestation. His thinking was that the more trees were cut down, the less carbon dioxide the land could absorb (this is where albedo plays its part), so more and more of it collected in the atmosphere. Wilson then remarked that the warming climate could bring severe drought to the Mediterranean, California, and Australia, but we shouldn't worry, because "the losses might be compensated for by improved conditions in places that are now wet and cold."[21]

*   *   *

More about the atmosphere: Lest the reader think that IGY was concerned only with high-profile, headline-grabbing research and discoveries, I'd like to end this chapter with a brief summary of the objectives and results from some of the lesser-known, smaller surprises of the IGY meteorological program, taken from the 1965 *Report on the U.S. Program for the International Geophysical Year*:

Summary of Meteorology Program

The meteorology program was designed to increase knowledge of the structure and motions of the atmosphere. The height of aerological soundings was increased, strategic new stations in the southern hemisphere were established, in cooperation with other countries in order to strengthen the 70°–80°W pole-to-pole chain; stations were established on drifting ice floes in the Arctic, geochemical, solar radiation and ozone measurements were made at many stations in the U.S. and in the Arctic and Antarctic.

South American Upper Air Observations

Objectives: This project of the U.S. Weather Bureau was organized in order to complete the North-South chain of stations along the 80° W meridian, in accordance with CSAGI recommendations and in cooperation with the South American Meteorological Service.

Results: Following the termination of the IGY, the upper-air programs at the five stations were conducted for almost a year by the U.S. Weather Bureau under sponsorship of the Air Weather Service of the Air Force in cooperation with the South American Meteorological Services. Subsequently, all programs, with the exception of that at Guayaquil, were continued by the respective governments concerned. The data collected during the IGY not only supported various research projects but contributed also to operational requirements of the various meteorological services and the requirements of air transport.

High Altitude Rawinsonde Observations

Objectives: This project, in accordance with CSAGI recommendations that efforts be made to secure upper air observations to 30,000 meters, provided for extra-performance balloons for U.S. upper air stations.

Operations: Independent of the IGY, balloon manufacturers were developing higher performance balloons. Air Force meteorologists had developed an 800-gram balloon which served as the basis, in 1955, of a recommendation that an equivalent balloon be used at all U.S. IGY stations. By the beginning of the IGY, however, the standard 600-gram balloon was improved.

Results: A significant result of this program is the preparation by the U.S. Weather Bureau Stratospheric Analysis Project of a series of 10-millibar charts covering the IGY period. This series is composed of thrice-monthly charts spaced at 10-day intervals, covering the region 10°W to 150°E and 10°-20°N to 90°N. In the first volume of the series, there are thirty-six charts drawn for the 1,200 GMT observation of the 5th, 15th and 25th of each month from July 1957 to June 1958 inclusive. The second volume extends the series from July 1958 through June 1959.

Geochemical Measurements

Objectives: This project of the U.S. Weather Bureau provided for measurements of carbon-dioxide and ozone, two elements present in the atmosphere in small amounts that are important factors in the atmospheric energy budget and may also be useful as tracers.

Results: The IGY ozone data obtained in the Antarctic have been studied in terms of the atmospheric circulation patterns. Wind and ozone roses were constructed to examine the variation of surface ozone with wind direction and in addition the net meridional transport of surface ozone was computed for each month for a year. Based upon these results a model for atmospheric circulation in the Antarctic has been proposed.

The CO_2 program has divulged a number of interesting features. First, a seasonal variation in the abundance is found in the Northern Hemisphere with a maximum in late spring or early summer and a minimum in late summer or autumn. No similar variation is as yet detectable in the Southern Hemisphere.

Interview with James Marshall Shepherd, PhD

Director, Atmospheric Sciences Program, University of Georgia

Mark O'Connell: The question I'm asking all my interviewees is this: If the IGY can convince six thousand scientists around the world to set aside their egos, their politics, and their personal and professional biases to solve some of the world's greatest scientific puzzles, can the scientists the world over do the same to tackle the climate crisis? Are they doing it already?

Marshall Shepherd: * I think we are. I think there's increasingly more multidisciplinary and convergent thinking. In trying to break down silos, as a meteorologist and atmosphere scientist, I now work with sociologists and oceanographers. I think IPCC (Intergovernmental Panel on Climate Change) is a good example of this legacy of IGY. IPCC is the big report that comes out every year that sort of assesses the global climate. It has won the Nobel Prize. By the way, we have a smaller version here in the United States called the National Climate Assessment (NCA). The first President Bush actually signed into law that the U.S. government has to assess the state of the climate every year for the United States. So I think these are different sorts of legacies from

---

* Dr. J. Marshall Shepherd is a leading international expert in weather and climate and is the Georgia Athletic Association Distinguished Professor of Geography and Atmospheric Sciences at the University of Georgia. Dr. Shepherd was the 2013 president of the American Meteorological Society (AMS), the nation's largest and oldest professional/science society in the atmospheric and related sciences. Dr. Shepherd serves as director of the University of Georgia's Atmospheric Sciences Program and full professor in the Department of Geography, where he was previously associate department head. Dr. Shepherd also holds a joint appointment in the College of Engineering and is the associate director for climate and outreach in its Institute for Resilient Infrastructure Systems (IRIS).

IGY that have spun off, that are sort of more science focused. Some of them were sort of grassroots-driven activities, you know; Earth Day, for example, or World Water Day. I think that the legacy of IGY is there.

M.O.: Okay, so you think scientists are coming together right now?

M.S.: I think we have no choice. And I think we saw it with things like COVID, too. I mean, there was conflict. We were faced with a very challenging crisis. And it's the same thing with climate. And I think we now understand and know that climate science, climate change, vaccines, COVID, those things are not going to be solved by [scientists in] independent individual silos, working in isolation.

M.O.: Where do you see the most room for improvement?

M.S.: I think those scientists and scholars—not just scientists, scholars in general—are trying to work across boundaries. Those boundaries are so entrenched, those silos are so entrenched that we still, at times, speak different languages, have different approaches and methodologies, and it can create challenges for working together, even though we know we need to.

M.O.: As an educator, you spend a lot of time with students, researchers, who I am assuming have a very positive and hopeful outlook on the work, and that they're pursuing that work because of that inherent optimism. How does that affect you to work with students?

M.S.: Yeah, students are vibrant, and they give you energy, and they're curious, and they want to learn. You know, I was at NASA for twelve years before I joined the university. They were both highly stimulating environments. But I think there's a vibrancy that when you have students around and bright ideas and just seeing light bulbs go off, and they discover something new in their own research or in a class, you know, it keeps you hopeful. Even as you hear a lot of those students, parents, and grandparents spout science based more on their ideologies and their belief systems than data and facts and objectivity. I think the current generation of students, for the most part, are moving us away from that.

CHAPTER 10

The Oceans Beneath

As we learned in the previous chapter, the sciences of oceanography and meteorology are closely tied together, and nowhere is this more evident than in the IGY program of the USSR.

One of the Soviet Union's largest, most ambitious IGY programs dealt with global oceanography and climatology, as related by Soviet oceanographer Artjob Povzner in an oral history interview recorded for AIP in 1999. "Yes, about the scientific problem of the oceanography," Povzner related in somewhat halting English, "this program in the USSR consists of thorough observation of the sea level and in variation of this sea level and oceans. Then long period waves, then the deep-water circulation."[1]

"During IGY they discovered a lot of [ocean] currents on different levels, different speeds, velocities. It was very interesting," Povzner continued. "And it was not possible in this measurement if you could not have the measurements and data in oceanic forms—it's called the native form—at the same time. For example, one vessel can measure this current here in order to determine the direction of this place of this current. And from this point of view it's the science, the oceanography, in general is also very important to combine this type of data with meteorological data to the same time. Meteorological data was very important for the study of the problems on the directions on the ocean surface and atmosphere. Here you had no problem that somebody refused to give data before they will be published because everybody was interested to have the data from the neighbors."[2]

As was the case with many IGY programs, global synchronization of tests, experiments, probes, and data collection performed by scientists or remote devices was the key to understanding how the world worked.

"We have to get the comparable results of the whole world," said Povzner. "It seems to me it was a first difficulty. The main difficulty. How to decide this problem, to build new stations, for example, in the places, in the cities where we had no observations, for example. To organize new expeditions in the world oceans. For example, in the beginning of IGY, you have only three oceanographic ships. During the IGY we could use about twenty ships because we could invite the other ships and other fishers [sic] and also military. And they worked for the IGY."[3]

"You cannot measure the currents without participation of other countries because you have too small a number of vessels or network of meteorological stations," Povzner added. "You need a thousand meteorological measurements. It's not about forecasting of the weather. No, I speak about science meteorology, like ozone. Scientific, physical assessment must be done. You need a lot of materials of data from different regions. You cannot solve the problems of atmospheric physics without using the data of many countries from different regions."[4]

Another IGY oceanographer, W. Arnold Finck, recalled in a 1997 oral history interview that IGY answered the prayers of many ocean researchers who struggled to secure funding for their pet projects. "I think the International Geophysical Year gave us the opportunity to explore many areas that we would like to have done, but it gave us the funds to do it."[5]

"Since it was a cooperative effort with many countries it was very successful," added Finck, who spent much of his career watching the oceans and what lies beneath them at the Lamont-Doherty Earth Observatory in Palisades, New York. "I think that probably one of the most interesting things that went on was the work of the research vessel *Vema* which was probably the only research vessel that just went out and circled the globe and was out all the time.

"It was always fun to see what they brought back because they'd always have some new discovery when they came back. In those days it was not uncommon to have some new ridge or some new mountain under the sea or something come up. It was fun."[6]

One IGY oceanographer took the fun of data collection to a new level. Charles David Keeling, a postdoctoral student at the California Institute of Technology, was known in scientific circles as a man who was incessantly measuring things—a bit like the ever-numbering Count on *Sesame Street*, it seems—and one of the things he enjoyed measuring the most was atmospheric CO_2. This was a natural fit, as Keeling was passionate about being outdoors and communing with nature, and measuring CO_2 in the atmosphere was as close to nature as he could get.

"Keeling's research program and his exacting personal standards pushed him to make measurements at a level of accuracy that no instrument on the market could reach," Spencer Weart wrote. "Keeling visited wilderness areas all around California, laboriously refining his techniques. He was surprised to find that at the most pristine locations he kept getting the same stable number—a true base level of atmospheric CO_2."[7]

Keeling's findings spurred new research into the greenhouse effect, the process by which gases like CO_2 trap heat in the atmosphere and over time raise the earth's surface temperature. Funding from IGY, as mentioned by Finck, allowed Keeling to purchase infrared gas analyzers to perform continuous measurements of CO_2 at the Little America Antarctic base and at the new observatory built on the volcanic peak of Mauna Loa in Hawaii—where some of the cleanest air can be found—to establish a global, reliable baseline CO_2 level. Keeling's idea was to reestablish the baseline every ten or twenty years to see whether and how much the baseline had changed. "To actually detect a rise of the CO_2 level during the 18-month term of the IGY," Weart wrote, "scarcely seemed possible."[8]

And yet that's exactly what happened.

*   *   *

"With a meticulous attention to detail that verged on the obsessive, Keeling managed to extract a remarkably accurate and consistent baseline number for the level of CO_2 in the atmosphere," Weart wrote. "In late 1958, his first full year of Antarctic data hinted that a rise had actually been detected."[9]

And still there was no alarm, no call to action. Keeling's findings were interesting, but as of yet not highly valued. While few (if any) scientists doubted that human use of fossil fuels was causing global warming, it was widely believed, and had been for some time, that the vast majority of the CO_2 we were creating was absorbed by the oceans, where it would settle to the seabed and do no real harm. The bottom of the ocean was already being bandied about as an ideal dumping ground for radioactive waste, as it was thought the water in the oceans takes hundreds of years, and maybe more, to "turn over." If the ocean floors were so stable and so safe for storing nuclear waste, many thought, they ought to be a safe repository for Earth's CO_2.

"A particularly simple and powerful argument was that the added gas would not linger in the air," Weart wrote. "Most of the CO_2 on the surface of the planet was not in the tenuous atmosphere, but dissolved in the huge mass of water in the oceans. Obviously, no matter how much more gas human activities might pour into the atmosphere, nearly all of it would wind up safely buried in the ocean depths."[10]

Or so we thought. By 1958, Keeling was conducting his research at the California-based Scripps Institute of Oceanography under the watchful eye of Roger Revelle, a scientist as fond of collaboration as Keeling was of measuring things. With a surge in Cold War scientific funding at his disposal and a wealth of experience analyzing seawater chemistry—including that of seawater irradiated by nuclear bomb testing on the Bikini Atoll—Revelle was inspired by the meticulous work of Keeling and others to conduct ongoing CO_2 monitoring in both the ocean and the atmosphere well beyond the span of the IGY. "Roger felt very strongly this was important to humanity," said Revelle's Scripps colleague Harmon Craig.[11]

In that, Revelle was right, but it's important to mention that his first discovery for IGY took place on a deep-sea exploratory mission on a research vessel owned by Scripps. As the result of some rather sophisticated magnetic seafloor probing and mapping—one of the primary objectives of the IGY oceanographic program—Revelle and his staff came to the rather astounding and unexpected realization that the seafloor was much younger than was supposed. It seems reasonable that the deeper we penetrate under the ocean floor, the older the rocks and the sediment will be, at least from a traditional archaeological perspective, in which the deepest items are always the oldest, but it was not so in this case. "Of course, this is really the basis of our modern understanding about the earth, that the sea floor is very young," Revelle said in his oral history. "It's renewed constantly by volcanic activity in the mid-ocean ridges, and then the volcanic rock spreads across the ocean and then is subducted in the trenches."[12]

Back on dry land after his fruitful oceanic mission, Revelle turned his attention to Keeling's CO_2 records from Antarctica and Mauna Loa, and he went right to work deciphering the data. "It was during the IGY that we started," Revelle recalled. "In the 1950's it was realized that we were producing quite a bit of CO_2 by burning fossil fuels, but it was thought that nearly all of it must be going into the ocean," Revelle said. "And the reason that was believed was that the ocean contains about fifty times as much CO_2 as the atmosphere, 35,000 gigatons in the ocean, about 700 gigatons in the atmosphere. And so it was thought that the partition of CO_2 would be in that ratio of 50 to 1, which would mean that the atmospheric carbon dioxide would be going up very slowly."[13]

"It turns out, from Keeling's measurements, and from predictions made by [radiocarbon dating pioneer Hans] Suess and me in 1957, that in fact about half the CO_2 released by the burning of fossil fuels would stay in the atmosphere," Revelle wrote. "This is because . . . if you increase CO_2 in the atmosphere by say one part per million, you increase the CO_2 in the ocean by only a tenth of a part per million, a big difference. And, that's because of the way that the carbon dioxide in the water is partitioned between carbonate ions and bicarbonate ions and free CO_2, and as you add a little CO_2 to it, it tips the equilibrium between these three kinds of carbon dioxide, so that the result of this buffer. . . . [I]t means that you can add quite a bit of CO_2 to the air without getting much in the water, and vice versa. So, that was started in 1957 and has been continued ever since."[14]

And there you have it: the discovery of global warming and the effects of that discovery on an unconcerned world.

*   *   *

More about the oceans: Lest the reader think that IGY was concerned only with high-profile, headline-grabbing research and discoveries, I'd like to end this chapter with a brief summary of the objectives and results from some of the lesser-known, smaller surprises of the IGY oceanography program, taken from the 1965 *Report on the U.S. Program for the International Geophysical Year.*

Summary of Program

The U.S. program in oceanography was devoted primarily to four categories of work: a) the study of changes in sea level ranging from wave action to seasonal variations; b) the study of the water of the deep oceans, particularly its circulation, and the geophysics of the ocean areas; c) the distribution of carbon dioxide in the sea and atmosphere and the use of radio-isotope trace elements to study oceanic circulation; and d) the study of the polar seas. In addition, two specialized projects were carried out, one to purchase bathythermographs for the research institutions, the other to process bathythermograph data.

Sea Level Studies, Pacific

Objectives: This project, organized at the Scripps Institution of Oceanography, La Jolla, California, for the study of sea level and its fluctuations, was proposed as an IGY project in order to increase our understanding of the water budget of the oceans. One of the fundamental problems is to ascertain whether there is actually a seasonal mass change in the oceans or if the sea level change can be accounted for by volumetric considerations. At the same time, it was desired to study short period changes in sea level covering the spectrum from ocean and storm waves, tsunamis, and fluctuations of several days period.

Results: The analysis of Tide Gauge and Steric data has confirmed the fact that sea level is high at the end of summer in each hemisphere. However, the improved distribution of samples during the IGY led to the following new information: a. Sea level variations in the South Pacific were measurably smaller than those in the North. b. The seasonal variation was appreciably larger near the continents than in the central oceanic regions of the Northern Hemisphere, but this effect was virtually absent in the Southern Hemisphere. c. The December–June oscillations found previously in the Gulf of Alaska can now be described in more detail. d. The tide gauge results are in good agreement with the ISOSTATIC variation computed from the steric data and the barometric pressure, i.e. the observed variations in mean sea level were in isostatic equilibrium over most of the Pacific Ocean.

Some 96 months of useful wave records were obtained during IGY from the various island stations, including records from two tsunamis and three hurricanes. By far the majority of anomalous day-to-day sea-level fluctuations were related to local barometric pressure disturbances. Three unusual records are shown in Fig. 1 [not included]. The upper record from Kona, Hawaii, shows waves produced by the high-altitude hydrogen bomb test of August 12, 1958, at Johnston Island, 750 miles to the southwest. The lower record contains the earliest portion of the wave series associated with the large tsunami of March 9, 1957, and is the first record of a tsunami ever made at a small island by an instrument specially designed for this purpose. The center record—at first, interpreted as waves emanating from a large but remote storm center in the north-central Pacific—was caused by failure of the hydraulic sensing hose leading over the reef at Wake Island into deep water. It clearly shows the difference in relative amplitude of swell when recorded at deep and shallow depths, respectively. In all three records the amplitude scale is roughly one-third of actual amplitude, as recorded on a chart five inches wide. Typical daily background precedes the onset of unusual activity on each record. This background is principally swell "noise" which clearly varied seasonally by a factor of ten between winter and summer in the north Pacific.

Oceanic Circulation and Geophysics, Pacific

Objectives: One of the major aspects of the IGY program was the study of oceanic circulation. Under this project at Scripps Institution of Oceanography, La Jolla, California, three separate cruises were organized. The first of these cruises included almost all facets of physical, chemical and biological oceanography and marine geology. The second and third cruises were devoted primarily to the detailed study of shallow currents in the Central Pacific.

Results: a. Expedition DOWNWIND. A preliminary report of this expedition was published by World Data Center A (General Report Series No. 2, June 26, 1958, "Expedition Down wind, University of California, Scripps Institution of Oceanography Cruise to the Southeast Pacific"). A complete description of the results of this expedition is beyond the scope of this report but several significant results may be mentioned.

Of extreme interest to students of the structure of the earth were 32 measurements of the heat flow through the ocean floor. This number of measurements almost equaled the total number taken previously in all oceans. It was found that the heat has stimulated some new ideas on crustal structure and tectonic phenomena. Several major features of the ocean floor were explored and mapped in detail, including the South American Trench, the Nasca Ridge, and the East Pacific Rise (also called the Albatross Plateau). Measurements of the concentration in the atmosphere of carbon dioxide revealed a striking uniformity in the concentration of this substance over a wide range of latitude.

b. Expedition DOLPHIN. John A. Knauss and Joseph E. King published a preliminary description of the findings of this expedition in *Nature*, vol. 182, Aug. 30, 1958, which confirmed the existence of the undercurrent as suggested by Cromwell. Knauss and King proposed to name the current the Cromwell current in honor of its discoverer who perished in an aircraft accident during the IGY. This current is one of the major features of oceanic circulation, with a length of about 6,000 km and with a transport of approximately 30 million cubic meters of water per second. The understanding of the role of this enormous current in the heat and water budget of the Central Pacific ocean is a new and important problem for the theoretical oceanographer. The results of the expedition were summarized by Knauss in *Deep Sea Research* (1960).

c. Expedition DOLDRUM. The Equatorial Countercurrent had been known previously and its extent and transport estimated by Sverdrup, this expedition, by direct observation, provided the unexpected information that, while the core of the current was shallow as had been previously suggested, the flow extends to considerably greater depth than estimated and in fact extends well below the thermocline and indeed most of the transport is below the thermocline. The new estimate of the transport is 50 million cubic meters of water per second or approximately two and one-half times the previous estimate. This current, like the Cromwell current, is one of the major circulation features of the Pacific Ocean.

Sea Level Studies, Atlantic

Objectives: This project was organized at Columbia University's Lamont Geological Observatory, Palisades, New York, for the study of sea level and its fluctuations, which was proposed as an IGY project in order to increase our understanding of the water budget of the oceans. One of the fundamental problems was to ascertain if there is actually a seasonal mass change in the oceans or whether the sea level change can be accounted for by volumetric considerations. At the same time, it was desired to study short period changes in sea level covering the spectrum from ocean and storm waves, tsunamis, and fluctuations of several days period. In addition, recording microbarographs and microvariobarographs, as well as short period wave recorders, were installed for the purpose of studying the interaction between the pressure fluctuations in the atmosphere and effects on the sea surface.

Results: An anomalous drop in sea level of .75 feet was observed in August 1959 at Bermuda. By September the sea level had returned to its July value. This change, it is thought, is probably related to a local variation in the vertical ocean structure. The barograph and wave instrumentation, on the other hand, have produced a considerable amount of information. From several cases examined during the IGY it appears feasible to forecast potentially damaging storm surges in the islands of the Lesser Antilles. Observations from the Texas Tower No. 4, eighty-five miles southeast of New York City, have provided important data for the study of direct energy coupling between atmospheric phenomena, such as the jet stream, and the sea surface. Atmospheric pressure records have given indications of gross atmospheric oscillations which are suggested to be of solar or tidal origin. Atmospheric waves from Soviet nuclear explosions which were detected by IGY microbarographs are providing the basis for a study of gross atmospheric structure.

Oceanographic and Geophysical Investigations—Atlantic

Objectives: Under the auspices of the Lamont Geological Observatory of Columbia University three major cruises of the Lamont vessel VEMA were devoted to IGY investigations to obtain a more complete scientific understanding of the geologic structure of continents and ocean basins, to ascertain the large scale circulation of the ocean waters, to study the nature and origin of the sediments on the ocean floor, and to learn more of the biota living in the ocean and on its floor.

Results: Comprehensive geophysical expeditions like those of VEMA produce a great mass of data that cannot be adequately summarized in a report of this nature[;] some individual findings, however, may be noted as examples of the work accomplished. During VEMA-12, extensive biological sampling of the deep ocean basin was undertaken. During this work, in which hundreds of specimens from great depths were obtained, two macro-organisms reached the surface alive. A small sand flea, a member of the amphipod group of crustaceans, and a worm, survived temperature changes of about 27°C and pressure changes of 6,600 and 8,100 pounds per sq. inch, respectively. The opportunity to study a living creature from oceanic depths of 13,000 to 16,000 feet is a rare treat for the marine biologist.

Bottom sampling on VEMA-14 revealed a thin layer of pebbles on the ocean bottom having no covering sediments, thus indicating a recent deposition. Biological sampling in the Peru-Chile trench produced during VEMA-15 four specimens of a class of Mollusca thought, until 1957, to have become extinct in the Devonian. The specimens are considered to represent a new subgenus and species: Neopilina (VEMA) ewingi and the discovery suggests that additional relict types may exist alive in the deep sea off Central and South America.

Since the introduction of the precision depth recorder with a short ping (about 5 millisec) sub-bottom reflections have been recorded in the deep sea. During VEMA-15 such an echo was observed over a large area of the eastern part of the tropical Pacific Ocean. A piston corer was used in these areas and a layer of clean white ash was encountered at the depth indicated from the PDR trace. The white ash was identified by D. B. Ericson of Lamont as a very fine-grained ash of clear glassy fragments, many of which are needle-like. Since the ash is fairly near the surface, is not discolored, and contains nothing but the glassy ash material, it is concluded that it was laid down fairly quickly. Foraminifera deposits beneath and above the layer afford a possibility of dating the deposit. The further exploration of this layer and study of the dates and variations in contaminants and internal structure should lead to important geophysical information.

Oceanographic Survey of the Atlantic Ocean

Objectives: The study of the large-scale circulation, particularly the deep currents, of the ocean was one of the prime objectives of the IGY program. A comprehensive survey of the Atlantic Ocean to further this objective was organized under the auspices of the Woods Hole Institution of Oceanography.

Results: The joint ATLANTIS–DISCOVERY I work near the coast of the United States was a striking success. It had been suspected by Henry Stommel of Woods Hole, as a result of theoretical considerations, that there should be a deep current running counter to the Gulf Stream in the western Atlantic. The Swallow floats tracked by DISCOVERY I confirmed that this current exists. A layer of virtually no motion was found between 1,500 and 2,300 meters. Above this layer the water moved to the northeast with expected velocities, below this layer motion was measured to the southwest, with velocities ranging between 2.6 and 9.5 cm/sec. At 2,800 meters depth velocity increased to between 9.7 and 17.4 cm/sec. The volumetric transport was computed to be between 3.5 and $5 \times 10^{\wedge}$ cubic meters/sec.

CO_2 Measurements and Radiochemical Analysis of Sea Water

Objectives: This project at the Department of Oceanography, University of Washington, was part of an integrated program to determine the average CO_2 atmospheric concentration and to study the exchange of CO_2 between sea water and the atmosphere. In addition, radiochemical analysis of sea water was undertaken to provide information on age of water masses, circulatory patterns, and geochemistry of marine processes. Radiochemistry techniques were also used to determine rate of primary production.

Results: Radiocarbon dating of North Pacific waters, when compared with results reported by Lamont workers, suggest that the North Pacific and Aleutian Trench deep waters may be as much as three times older than the North Atlantic deep water. A more reliable comparison will result from the further analysis of Pacific deep and surface water samples. The thirty-four radiocarbon samples collected under this project have a distribution, both in space and time, that should provide valuable information once highly precise radiocarbon measurements are completed.

Interview with Wallace J. Nichols, PhD

Research Associate at California Academy of Sciences

Mark O'Connell: You recently described to me that you had submitted a research paper on the topic of sea turtles to a scientific journal, and that there were three hundred names, including yours, listed as authors of this paper. That's quite a collaborative effort! Can you tell me more about that?

Wallace Nichols: The paper is submitted and fingers are crossed. Journal articles can be especially, incredibly unwieldy, especially when multiauthored. Ambitious submissions. It's a feat. I mean, really the organizing of all that data and all those coauthors, it's a quite a feat. So we have our fingers crossed; there are three hundred of us with our fingers crossed.

M.O.: How has IGY affected your field? I know that with a lot of fields of science involved in IGY, they'll have one huge claim to fame, like, "This is what we discovered through IGY." And I know with oceans, we were able to measure the ocean floors for the first time. What was the big, big accomplishment for IGY?

W.N.: I'd say, a recognition that there's a lot to learn, a lot to explore, the underwater mountain ranges of formidable size, and canyons and just vast, unknown terrains. You know, riverine valleys that are unseen. But when you get into the bathymetry, you see the history of sea level rise and fall.

Also, for oceanographers, I would say that they've led the way in terms of mass collaboration. And I think that that kind of kicked the door open, and perhaps kicked it off the hinges, not just open, to never be closed again: the recognition that the art of collaboration is absolutely critical to studying the ocean. Because you've got all these nations and their oceanographic institutes and their government agencies and their own limited reach, or vast reach, as the case may be with the UK and France and the United States. But you're not going to understand the ocean, the contiguous global ocean, without a massive amount of collaboration. And I can say, as a biologist, we've lagged.

Sea turtle biologists are incredibly collaborative, as a culture. In contrast, marine mammal biologists are wildly competitive and need to learn a few things. Which has been a setback. So you start seeing oceanographic publications with these massive coauthor lists. And it's the norm, and then you move almost in any direction in science, and you'd find that to be much less prevalent. Maybe in some realms of physics. But really, the art of collaboration was really, I would say, defined by the effort that you're writing about, and we're not there yet. There's still quite an academia, and you know, in science there's still a lot of "terrainism" or competition, professional jealousy. Because you'll get, "No, it's my idea. I had this idea first," rather than these big, massive, coauthored efforts.

M.O.: Back to the ocean mapping. Was there any sort of new technology or new field of thinking that allowed us to map the ocean floors for the first time during it? Why?

W.N.: What generally occurs with oceanography is that the military technologies tend to drive the train. And usually, it's not the academic institutions that have the budget to invest. Now this is changing. And we're in a new, completely new era of billionaire philanthropist explorers. So you've got the likes of James Cameron, and Richard Branson, and Elon Musk, and Bezos. And if Steve Jobs were alive, he might be fascinated by the ocean and big science with his wealth. For a while there, the billionaires were all building submarines to go explore the deep ocean, and now they're all building spaceships. And so back in the day you had the billionaires building libraries and lovely parks or institutions on land, but not exploring as they are now. So I think that's kind of a shift. You had a military interest in submersible submarines, communication, underwater defense mechanisms that drove the technology and sort of allowed for some of these, say, ancillary explorations to occur. Map the seafloor for scientific reasons. But the technology would probably best be described as driven by security. Security and perhaps aggression.

M.O.: My biggest surprise in the research I've done for this book has been the discovery that back in 1957, some scientists were aware that the earth was getting warmer, that the atmosphere was getting warmer, that the oceans were getting warmer. They were aware. I was just staggered. It's come up over and over again: "Oh, by the way. The earth is getting warmer. And maybe someday that'll be a problem. But not today." Have you encountered this?

W.N.: I think that it's only in the last decade that it's become a combination of fashionable and politically charged. Maybe it's a little longer than that, but fifteen to twenty years at the most. And when I was a graduate student, I was at Duke University and worked on a report on sea level rise. That was 1991–1992. And we concluded that yes, sea level is rising, going to rise. And we should be planning for it. And at that time, the connection between what we now call climate change and the ocean was very weak, it was not really considered.

It's really recent that the ocean-climate connection has been expounded upon and really developed in terms of the minute science was there, but it was just not really discussed all that much. And so I would say, looking way back, you have scientists who are paying attention to the signals and thinking about, you know, the projections of where this may be going and why it may be occurring. And concluding that, yeah, we're, we're onto this warming trend.

M.O.: It could be described as a failure of imagination.

W.N.: I think we are constantly failing to imagine, in all realms constantly. I think apex leadership is rare, exceedingly rare. Managerial leadership is common and useful. But apex leadership, the kind where you, you have an idea that you have reason to believe is true, that nobody else is really having at the time that you're having it and you look around, and it's just you at the tip of the spear. But you move forward with it. Anyway, I think a lot of people look out both sides of their eyes and make a decision to not move forward until society or their peers catch up. Or the funding catches up, but the combination of failure of imagination and lack of apex leadership combined leads us to hang on tight to the status quo.

M.O.: I've been really curious since our first contact to hear more about your interest in neuropsychology; you had described it as the hardest science but the most important area of research for oceanography. I would love to hear more about that.

W.N.: Yeah, the fifty-thousand-foot view is that when we undervalue anyone, or anything, bad shit happens. History has shown that when we undervalue the ocean, or the lakes or the rivers or the forests, we wreck them. And then we say, well, wait a second, why do we do that? Why is our river dead? Why is our forest gone? Why is the ocean in such bad shape? Why are there no fish? Why are there dead zones? And part of the reason is we have undervalued nature. Likewise, history has been clear that when we undervalue nature, we tend to make poor decisions.

And so the value equation in it is flawed. And a big reason is that we have dialed out the emotional well-being benefits of a healthy forest, a healthy river, healthy ocean. We've said keep that stuff out. It's soft. It's not science, it's subjective. It's not quantitative. It's hippie dippie, woo-woo, eyeball-rolling stuff. Turns out that's wrong. Turns out that healthy nature is really good for our cognitive, emotional, psychological, social, spiritual, and physical well-being. Research is increasingly clear on that. We're getting there. That's part of what I do every waking minute. And so the idea is how do we fix, how do we update the value equation to reflect this idea that healthy waterways, not just oceans, but water and all of its forms can not only provide economic and ecological benefits but vast emotional well-being benefits to all people? And so in our society, the way you convince the powers that be that that is true is science. Every spiritual tradition, every culture, every sacred text, says that water is medicine for your spirit.

But now we're coming back and saying, wow, we've got a water pollution crisis, a biodiversity crisis, a climate crisis, and a mental health crisis on top of it all, and each exacerbates the others. So if you're looking for a big lever to come at all of those crises, simultaneous existential crises, connecting well-being to water, the basis of all life on this little planet is maybe a pretty good one. That's a pretty big lever. It tends to cut through political divisiveness quite well. And because water is the single most important thing about life on Earth, and the most complex thing in the universe is the human brain, the mind, when you put those two realms together, it's really interesting. And it's the hardest science. Unless you're at an astrophysics conference, and then maybe it's kind of number two.

M.O.: Are we too late?

W.N.: I know I vacillate.

M.O.: That's an honest answer.

W.N.: Some weeks or days, full of hope and optimism. One of the sources of optimism would be that this conversation that we are having has so much potential. There's this ripe, pregnant moment that we're in. And I think the approach that we've taken is sufficiently swarmlike. There's just a lot of things going on in a lot of places, so not by design, more movement than entity oriented. We're getting there. The word is out.

CHAPTER 11

Aftermath

Here's the bottom line: IGY worked.

After all the money spent, the miles traveled, the rockets and balloons launched, the observatories and stations built, the discoveries made, the glaciers traversed, the samples taken, the sunspots counted, the measurements made, the meetings held, the data gathered, the observations conducted: IGY worked. What's more, it worked better than anyone could have imagined.

Not every IGY program yielded positive results, naturally, and many were not expected to yield either positive or negative results, since they were conceived as dedicated, information-gathering data collection programs, pure and simple. Nonetheless, all data produced by every IGY scientist the world over was equally valuable, as demonstrated by the mission statement of the IGY World Data Center (WDC) System: "Data constitute the raw material of scientific understanding. The World Data Center system works to guarantee access to solar, geophysical and related environmental data. It serves the whole scientific community by assembling, scrutinizing, organizing and disseminating data and information."[1]

In late 1958, as IGY started to wind down its research and exploration activities and take stock of the massive amounts of data it had generated, focus shifted to the creation and operation of data management plans for each of the scientific disciplines involved. The *Guide to*

Data Exchange quoted here was a 1996 update, just one of several updates over the years that kept the centers operating at peak efficiency, preserving the past and preparing for the future. "The IGY planners were remarkably prescient," the guide pointed out. "The 1955 recommendation mentioned that Data Centers should be prepared to handle data in machine-readable form, which at that time meant punched cards and punched tape."[2]

"The IGY was a tremendous success. The newly developed spaceflight capability was used to discover and explore Earth's radiation belts, to study the magnetosphere, and to provide the first observations of the emission from the Sun's corona," wrote Joseph M. Davila, Arthur I. Poland, and Richard Harrison in a 2001 article proposing an IGY follow-up called the International Heliophysical Year. "Public interest in the scientific results of the IGY was high. The IGY provided a forum and a backdrop for discussing the importance of geospace influences on Earth."[3]

That discussion was dramatically and substantially altered as IGY wound down and scientists returned to their home countries, laboratories, and classrooms filled with new thoughts and ideas. "There is nothing man is closer to than the earth beneath his feet," wrote *Life* magazine, "and hardly anything he has known less about. But even scientists, who were aware of their enormous ignorance of the earth's crust, were astounded at how much they had not known about it—and how much they learned about it during IGY."[4]

When asked in an oral history interview in 2001 what aspect of working on IGY made the strongest impression on him, ionospheric physicist and committee secretary Nathaniel Gerson recalled that he was inspired by the caliber of the scientists assembled by the leaders of IGY. Being responsible for taking the minutes at the IGY meetings, Gerson absorbed everything that was said and done by committee members. What impressed him the most was "seeing new people and being impressed with their grasp not only of atmospheric physics and geophysics but also their discussions about individuals and national policies and attitudes. Many had ties with foreign collaborators and knew second hand some of the objectives and motivations of the countries involved. The entire experience educated me to a new dimension of research: how to work with various groups, how to motivate them, how to think of the planet as a whole, and some of the future problems facing mankind."[5]

In the same interview, Gerson praised the leadership of Hugh Odishaw, director of the U.S. IGY program. "Skilled, polished, knowledgeable. I can't praise him too highly," Gerson said. "I think, as I look back, he had one overriding objective: to make it work, make it work well for the United States and internationally. Because the U.S. needed it. Weather is not just national. It doesn't stop at the boundaries. It's not just that it doesn't stop at the boundaries, you have to know what's going down. It's one earth. The U.S. has to know what's going on. What happens there can happen here. . . . As a principal power, the U.S. has an awesome responsibility. It must act correctly and from current, correct knowledge."[6]

"I think that there was substantially more science accomplished because of the cooperative efforts than would have been the case, if each country went its own private way," said physicist Herbert Friedman, PhD, of IGY in a 1983 oral history interview. "It also had great educational value. It encouraged more support for geophysical sciences from the responsible agencies. The original commitment of Eisenhower to support the IGY was something of the order of $11-million, I think, out of NSF funding. By the time the program was completed, the contributions were closer to, maybe $45-million for the U.S. effort in those two years, 1957 and 1958."[7]

"Because the IGY planning started several years before Sputnik [in 1957], you might say that the scientific community had developed substantial plans for what to do with satellites from a scientific standpoint. I think the establishment of NASA went ahead much more smoothly because the scientific community could project a very interesting and large scientific program. NASA's total commitment initially was to science. It is true that there were speculations about great practical returns, the ability to conduct radio communications via satellite was obviously bound to have a very large pay-off, and the ability to set up weather satellites and photograph the cloud cover of the earth was going to have a large pay-off for applications, public uses," Friedman said. "But the scientific projections were maybe even more persuasive publicly, because, immediately, there was the discovery of the Van-Allen Belts, which was a real spectacle, and there was a large amount of glamor and adventure associated with rocket shoots all over the world from shipboard, and rockets sent on balloons; and then two-stage rockets put up in all sorts of outlandish places."[8]

Friedman pointed out that some of the high points of IGY came about in the years after 1958: "There has been a constant honing of the techniques of doing global geophysical research, and doing it on an international basis. There have been several major programs conducted since the IGY."[9]

Friedman also recalled that in 1970, NASA was planning to launch a series of satellites to study the earth's magnetosphere, only to discover that the Soviet and the Japanese space agencies were planning to do the same thing. One of the challenges involved in studying the magnetosphere was that it was "time variable," according to Friedman. The only way to gather meaningful data about the magnetosphere was to make simultaneous measurements from different locations within the magnetosphere itself. Over time, the United States, the Soviet Union, Japan, and the European Space Agency (ESA) joined forces, putting a total of eight satellites into orbit to conduct studies of the magnetosphere, accompanied by thirty more satellites specially equipped to provide additional data-gathering support. "It is that kind of activity," Friedman declared, "that has developed since the IGY, and has been very effective for global studies.

"I could mention a number of others, but that gives us the clue to how we want to proceed from here on. First of all, the IGY was planned for one or two years. We now know that many of the major geophysical problems have time scales of 10, 20 or even longer lengths of time. So, it makes very little sense to plan programs which involve a great deal of effort and coordination to run only a couple of years.

We want them to run for one or two or more decades. We want to set up the kind of organization that will have that sort of longevity. That means we have to identify lead agencies in the United States to match the international agencies for continuing such activities over, say, at least one generation of scientific life."[10]

"I would hope that individual scientific groups in their respective countries would carry on their own creative efforts," Friedman concluded. "They would have enough imagination to think of new approaches, develop new instrumentation, and then try to fit them into the broader participation."[11]

It is difficult to follow up on Friedman's passionate in-person account of the end of IGY, unless it's with even more heartfelt eyewitness accounts. For that, I must turn to the five authors whose histories of IGY written between 1958 and 1961 gave me the foundation for my research. Each has his own unique take on the aftermath of what was at the time the most massive mobilization of scientists for a common cause in history.

Ronald Fraser wrote three years after the end of IGY:

> What is to be gained from the gigantic enterprise of the International Geophysical Year, uniting as it does the thought and effort of the scientists of fifty-four countries, costing as it must millions of pounds, involving, inevitably the setting up of over a thousand observational stations in the Arctic, in the Antarctic, or athwart the Equator? . . . [T]o anyone who has seen the whole development of the enterprise from its modest beginnings back in 1950, one of its most shining facets has been the practical demonstration of the truth that, given a certain attainable threshold of mutual interest, men of all nations are willing, nay eager, to join in a drive towards the same goal. Not merely to discuss the matter, but to do something about it.[12]

Walter Sullivan stated:

> The IGY promoted a revival of science in former colonies (where observatories had closed when the colonists went

home) and in other underdeveloped areas; it stimulated an interdisciplinary approach in which men trained in different, though related, fields worked together to resolve problems beyond the reach of any one of them. Finally, the IGY set in motion a number of research programs that became permanent. It seems likely that, as the world grows up economically, politically, and scientifically, there will be even more ambitious international undertakings.[13]

Alexander Marshack pointed out:

> The full results of the IGY will not be known for at least a decade. The results of satellite research will be accumulated over many years, after many launches. The patterns of space and spaceflights must be uncovered step by step. Scientists, engineers and technicians must be trained.
>
> But the *earth* is still the home of man. The storms, quakes, droughts and other processes that occur here will remain our principal concern. With knowledge of the scientific problems to be solved and an understanding of the difficult paths that have been followed so far, we can see more clearly into the future.[14]

Sydney Chapman wrote:

> At some time in the future a new intensified international scientific enterprise may be undertaken, perhaps far transcending the International Geophysical Year in range and power. At that time the remarkable success of the IGY will be an inspiration, and a guide, to the new effort.[15]

J. Tuzo Wilson noted:

> It may be that minds trained to seek truth in science will seek it in other fields also, and that in their increasing number may lie hope for the international co-operation essential to our future. During the IGY scientists the world over worked together for common ends and in so doing realized

that they held in their hands, besides the keys to the nature of the earth, the keys to the future of mankind.[16]

"May I add a few words?" inquired oceanographer and Soviet IGY program administrator Lilia Morozovskia in the course of her American Institute of Physics oral history interview in 1999. "Everybody was very enthusiastic, everybody. No matter how small a country, but always representatives were at an IGY conference. Always somebody is writing from all over the world. I think that IGY—of course I'm a patriot of IGY—but I think that IGY somehow broke this War between Russia and the Western world. Well, at least it made a great—how do you call it? Breakthrough. Because scientists have shown that we can trust each other. At least when it means our health, the health of our children, the health of our planet."[17]

Notes

Prologue

1. Walter Sullivan, *Assault on the Unknown* (New York: McGraw-Hill, 1961), 4.
2. Tim St. Onge, *Scientist of the Seas: The Legacy of Matthew Fontaine Maury* (blog), U.S. Library of Congress, July 25, 2018, https://blogs.loc.gov/maps/2018/07/scientist-of-the-seas-the-legacy-of-matthew-fontaine-maury.
3. M. F. Maury, *The Physical Geography of the Sea* (New York: Harper, 1855), xiii.

Chapter 1

1. J. A. Fleming, *The Proposed Second International Polar Year, 1932–1933* (Abingdon, UK: Taylor & Francis, 1932), 131.
2. Walter Sullivan, *Assault on the Unknown* (New York: McGraw-Hill, 1961), 16.
3. United States National Committee, *Outline of International Geophysical Year Program* (Washington, DC: National Academy of Sciences, 1955), 1.
4. Sydney Chapman, *IGY: Year of Discovery* (Ann Arbor: University of Michigan Press, 1959), 102.
5. J. Tuzo Wilson, *IGY: The Year of the New Moons* (New York: Alfred A. Knopf, 1961), 6.
6. Sullivan, *Assault on the Unknown*, 27.
7. Wilson, *IGY*, 39.

Chapter 2

1. Ronald Fraser, *Once Around the Sun* (New York: Macmillan, 1957), 97.

2. Ibid., 102.

3. *Report on the US Program for the International Geophysical Year* (Washington, DC: National Academy of Sciences, n.d.), ix.

4. Hugh L. Dryden, *The International Geophysical Year* (Washington, DC: National Geographic Society, February 1956), 298.

5. Marcel Nicolet, *The International Geophysical Year (1957–1958): Great Achievements and Minor Obstacles* (New York: Annals of the Institute of Radio Engineers, 1984), 303.

6. Hugh L. Dryden, "Scientists of Forty-Six Nations Study the Land, Sea, and Air All Around Us," letter, *National Geographic Magazine*, February 1956, 285.

7. Ibid.

8. L. V. Berkner, *The International Geophysical Year: Proceedings of the Institute of Radio Engineers (IRE)* (New York: Institute of Radio Engineers, 1958), 133.

9. Columbia Climate School, The Earth Institute, 2014, https://www.earth.columbia.edu/articles/view/3190.

10. Ibid.

11. Ibid.

12. Ibid.

13. Interview of Kenneth Hunkins by Mike Sfraga, April 11, 1998, Niels Bohr Library & Archives, American Institute of Physics, College Park, MD, www.aip.org/history-programs/niels-bohr-library/oral-histories/22591-3.

14. Berkner, *International Geophysical Year*, 133.

15. Sydney Chapman, *IGY: Year of Discovery* (Ann Arbor: University of Michigan Press, 1959), 107.

16. Alexander Marshack, foreword to *The World in Space* (New York: Dell, 1958).

17. Walter Sullivan, *Assault on the Unknown* (New York: McGraw-Hill, 1961), 37.

18. Ibid.

19. Ibid.

20. Nicolet, *International Geophysical Year (1957–1958)*, 305.

Chapter 3

1. J. Tuzo Wilson, *IGY: The Year of the New Moons* (New York: Alfred A. Knopf, 1961), 21.

2. Alexander Marshack, *The World in Space* (New York: Dell, 1958), 159.

3. "New Portrait of Our Planet: What IGY Taught Us," *Life*, November 28, 1960.

4. Ibid., 66.

5. Marshack, *The World in Space*, 154.

6. "New Portrait of Our Planet: What IGY Taught Us," *Life*, November 28, 1960.

7. Marshack, *World in Space*, 154.

8. Louise Lerner, "Eugene Parker, 'Legendary Figure' in Solar Science and Namesake of Parker Solar Probe, 1927–2022," *UChicago News*, March 16, 2022, https://news.uchicago.edu/story/eugene-parker-legendary-figure-solar-science-and-namesake-parker-solar-probe-1927-2022.

9. Louise Lerner, "The Solar Wind, Explained," *UChicago News*, https://news.uchicago.edu/explainer/what-is-solar-wind.

Chapter 4

1. J. Tuzo Wilson, *IGY: The Year of the New Moons* (New York: Alfred A. Knopf, 1961), 288.

2. "New Portrait of Our Planet: What IGY Taught Us," *Life*, November 21, 1960.

3. *Report on the U.S. Program for the International Geophysical Year: July 1, 1957–December 31, 1958* (Washington, DC: The National Academies Press, 1965), 855.

4. "New Portrait of Our Planet: What IGY Taught Us," *Life*, November 21, 1960.

5. Ibid.

6. Interview of Charles Bentley by Will Thomas, August 7, 2008, Niels Bohr Library & Archives, American Institute of Physics, College Park, MD, www.aip.org/history-programs/niels-bohr-library/oral-histories/33888-2.

7. Interview of Charles Bentley by Will Thomas, August 6, 2008, Niels Bohr Library & Archives, American Institute of Physics, College Park, MD, www.aip.org/history-programs/niels-bohr-library/oral-histories/33888-1.

8. Ibid.

9. Ibid.

10. Ibid.

11. Ibid.

12. Ibid.

13. Ibid.

14. Ibid.

15. *Conference on Antarctica Final Act* (Washington, DC: U.S. Department of State, December 1, 1959), 14.

16. "New Portrait of Our Planet: What IGY Taught Us," *Life*, November 21, 1960.

17. The Antarctic Treaty, December 1, 1959, https://www.ats.aq/e/antarctic treaty.html.

18. Walter Sullivan, *Assault on the Unknown* (New York: McGraw-Hill, 1961), 414.

19. Ibid., 414, 415.

20. "New Portrait of Our Planet: What IGY Taught Us," *Life*, 85.

21. Antarctic Treaty, Articles I–V.

Chapter 5

1. Alexander Marshack, *The World in Space* (New York: Dell Books, 1958), 107.

2. Ibid.

3. J. Tuzo Wilson, *IGY: The Year of the New Moons* (New York: Alfred A. Knopf, 1961), 131.

4. Melanie Windridge, *Aurora: In Search of the Northern Lights* (London: William Collins, 2016), 24.

5. Marshack, *World in Space*, 106.

6. Ibid.

7. Sydney Chapman, *IGY: Year of Discovery* (Ann Arbor: University of Michigan Press, 1960), 68.

8. "The Carrington Event—the Most Intense Solar Storm in Recorded History!" South African National Space Agency, November 17, 2021, https://www.sansa.org.za/2021/11/the-carrington-event-the-most-intense-solar-storm-in-recorded-history/.

9. Dr. Ryan French, *The Sun—Beginner's Guide to Our Local Star* (Glasgow: Collins, 2023), 57.

10. Windridge, *Aurora*, 175.

11. Alexander Marshack, *The World in Space* (New York: Dell, 1958), 108.

12. "What Causes the Northern Lights?" Royal Museums Greenwich, https://www.rmg.co.uk/stories/topics/what-causes-northern-lights-aurora-borealis-explained/.

13. Chapman, *IGY*, 68.

14. Ibid., 70.

15. Walter Sullivan, *Assault on the Unknown* (New York: McGraw-Hill, 1961), 212.

16. Ibid.

17. Wilson, *IGY*, 139.

18. Marshack, *World in Space*, 112.

19. Sullivan, *Assault on the Unknown*, 212.

20. Carl Gartlein, *Auroral Data Center Newsletter*, August 23, 1959, 213.

21. Marshack, *World in Space*, 107.

22. Chapman, *IGY*, 70.

23. Marshack, *World in Space*, 113.

24. Interview of Herbert Friedman by Martin Harwit, June 7, 1983, Niels Bohr Library & Archives, American Institute of Physics, College Park, MD, www.aip.org/history-programs/niels-bohr-library/oral-histories/28186.

Chapter 6

1. Anne Mikolajcik and Maggi Glasscoe, *History of Plate Tectonics* (Los Angeles: U.S. Geological Survey, Southern California Earthquake Center, 1998), http://scecinfo.usc.edu/education/k12/learn/plate2.htm.

2. Interview of Brent Dalrymple by David Zierler, June 7, 2021, Niels Bohr Library & Archives, American Institute of Physics, College Park, MD, www.aip.org/history-programs/niels-bohr-library/oral-histories/46964.

3. Ibid.

4. J. Tuzo Wilson, *IGY: The Year of the New Moons* (New York: Alfred A. Knopf, 1961), 224.

5. Gordon F. West, Ron M. Farquhar, George D. Garland, Henry C. Halls, Lawrence W. Morley, and R. Don Russel, "John Tuzo Wilson: A Man Who Moved Mountains," *Canadian Journal of Earth Sciences* 51 (January 2014).

6. "Geological Mystery Explained," UMKC Today Archives, https://info.umkc.edu/news/geological mystery-explained/.

7. West et al., "John Tuzo Wilson."

8. Ibid.

9. Geological Society, "Pioneers of Plate Tectonics: Harry Hess," https://www.geolsoc.org.uk/Plate-Tectonics/Chap1-Pioneers-of-Plate-Tectonics/Harry-Hess.

10. Ronald E. Doel, "The Earth Sciences and Geophysics," in *Science in the Twentieth Century* (Reading, UK: Harwood Academic Publishers, 1997), 394.

11. "New Portrait of Our Planet: What IGY Taught Us," *Life*, November 7, 1960.

12. Ibid.

13. Ibid.

14. Wilson, *IGY*, 273.

15. West et al., "John Tuzo Wilson."

16. Doel, "Earth Sciences and Geophysics," 395.

17. "UMKC Today," in *Geological Mystery Explained* (Kansas City: University of Missouri Press, 2013), https://info.umkc.edu/news/geological-mystery-explained/.

18. Lynn Sykes, "A Seismologist Present at the Discovery of Plate Tectonics," interview by Kevin Krajick, June 6, 2019, Columbia Climate School, https://news.climate.columbia.edu/2019/06/06/50-years-of-earth-shaking-events/.

19. West et al., "John Tuzo Wilson."

20. Doel, "Earth Sciences and Geophysics," 394.

21. West et al., "John Tuzo Wilson."

22. Sykes, "Seismologist Present at the Discovery of Plate Tectonics."

23. Dalrymple interview.

24. Ibid.

25. Ibid.

26. Ibid.

27. Ibid.

28. Ibid.

29. Interview of Charles Drake by Ronald Doel, May 22, 1996, Niels Bohr Library & Archives, American Institute of Physics, College Park, MD, www.aip.org/history-programs/niels-bohr-library/oral-histories/22583-2.

Chapter 7

1. White House Press Secretary James C. Hagerty, press release, July 29, 1955.

2. Walter Sullivan, *Assault on the Unknown* (New York: McGraw-Hill, 1961), 85.

3. Ibid., 66.

4. Ibid.

5. Ibid., 49.

6. E. Nelson Hayes, *Trackers of the Skies* (Cambridge, MA: Howard A. Doyle, 1968), 76.

7. Ibid., 25.

8. Ibid., 37.

9. Interview of Herbert Friedman by Richard F. Hirsh, August 21, 1980, Niels Bohr Library & Archives, American Institute of Physics, College Park, MD, www.aip.org/history-programs/niels-bohr-library/oral-histories/4613.

10. Sullivan, *Assault on the Unknown*, 72.

Chapter 8

1. Interview of James Van Allen by Brian Shoemaker (1997), Byrd Polar Research Center Archival Program, Ohio State University Knowledge Bank, posted August 1, 2008, https://kb.osu.edu/items/18f37f5d-9a39-5392-870d-0f63d2af2195.

2. Ibid.

3. Ibid.

4. Ibid.

5. Ibid.

6. Ibid.

7. Walter Sullivan, *Assault on the Unknown* (New York: McGraw-Hill, 1961), 118.

8. "The Father of Space Science," *Biography*, University of Iowa, Department of Physics and Astronomy, 2004, https://vanallen.physics.uiowa.edu/biography.

9. Van Allen interview.

10. Ibid.

11. Interview of Herbert Friedman by Martin Harwit, June 7, 1983, Niels Bohr Library & Archives, American Institute of Physics, College Park, MD, www.aip.org/history-programs/niels-bohr-library/oral-histories/28186.

12. "Father of Space Science."

13. "James Van Allen: Flights of Discovery," posted November 2, 2010, by University of Iowa, YouTube, https://www.youtube.com/watch?v=Fij6E1ZdAKs.

14. Doraswamy Venkatesan, "Genesis of the Van Allen Radiation Belts," *Johns Hopkins APL Technical Diges* 6, no. 1 (1983): 103.

15. Ibid.

16. James A. Van Allen, *Origins of Magnetospheric Physics* (Iowa City: University of Iowa Press, 2004), 6.

17. Ibid., 6, 7.

18. "The Sun's Awesome Impact," *Life*, November 28, 1960.

19. "What Are the Van Allen Belts and Why Do They Matter?" NASA, https://science.nasa.gov/biological-physical/stories/van-allen-belts/.

20. Ibid.

21. George H. Ludwig, PhD, Explorer Instrument Maker, U.S. Space Physics Data Archive, in "James Van Allen: Flights of Discovery."

22. "Scoping the Problem," in *Radiation and the International Space Station: Recommendations to Reduce Risk* (Washington, DC: National Academy Press, 2000), 7.

Chapter 9

1. Interview of James Marshall Shepherd by author, July 15, 2024. Subsequent quotations from Dr. Shepherd are from the same interview(s).

2. Interview of Wallace J. Nichols by author, May 1, 2024. Subsequent quotations from Dr. Nichols are from the same interview(s).

3. "New Portrait of Our Planet: What IGY Taught Us," *Life*, November 14, 1960.

4. Ibid.

5. Ronald Fraser, *Once Round the Sun* (London: The Scientific Book Club, 1958), 37.

6. Walter Sullivan, *Assault on the Unknown* (New York: McGraw-Hill, 1961), 288.

7. Alexander Marshack, *The World in Space* (New York: Dell, 1958), 71.

8. "New Portrait of Our Planet."

9. Ronald E. Doel, "Why Value History?" *Eos* 83, no. 47 (November 19, 2002): 2.

10. Ibid., 2.

11. "Albedo and Climate," University Corporation for Atmospheric Research (UCAR) Center for Science Education, https://scied.ucar.edu/learning-zone/how-climate-works/albedo-and-climate.

12. "Ice Albedo: Bright White Reflects Light," Scientific Visualization Studio, NASA, February 5, 2004, https://svs.gsfc.nasa.gov/20022/.

13. *Report on the U.S. Program for the International Geophysical Year: July 1, 1957–December 31, 1958* (Washington, DC: National Academies Press, 1965), 349.

14. Ibid.

15. J. Tuzo Wilson, *The Year of the New Moons* (New York: Alfred J. Knopf, 1961), 273.

16. *Report on the IGY*, 349.

17. Ibid., 350.

18. Spencer Weart, "Money for Keeling: Monitoring CO_2 Levels, in *The Discovery of Global Warming*," April 2024, https://history.aip.org/climate/Kfunds.htm.

19. Ibid.
20. Wilson, *Year of the New Moons*, 311.
21. Ibid.

Chapter 10

1. Interview of Artjob Povzner and Lilia Morozovskia by Tanya Levin, January 19, 1999, Niels Bohr Library & Archives, American Institute of Physics, College Park, MD, www.aip.org/history-programs/niels-bohr-library/oral -histories/33699-1.
2. Ibid.
3. Ibid.
4. Ibid.
5. Interview of W. Arnold Finck by Tonya Levin, June 13, 1997, Niels Bohr Library & Archives, American Institute of Physics, College Park, MD, www.aip .org/history-programs/niels-bohr-library/oral-histories/6948-3.
6. Ibid.
7. Spencer Weart, "Roger Revelle's Discovery," in *The Discovery of Global Warming*, August 2021, https://history.aip.org/climate/Revelle.htm.
8. Ibid.
9. Ibid.
10. Ibid.
11. Interview of Harmon Craig by Spencer Weart, April 29, 1996, Niels Bohr Library & Archives, American Institute of Physics, College Park, MD, www.aip.org/history-programs/niels-bohr-library/oral-histories/32508.
12. Interview of Roger Revelle by Earl Droessler, February 3, 1989, Niels Bohr Library & Archives, American Institute of Physics, College Park, MD, www.aip.org/history-programs/niels-bohr-library/oral-histories/5051.
13. Ibid.
14. Ibid.

Chapter 11

1. *Guide to the World Data Center System*, April 1996, https://www.ukssdc .ac.uk/wdc/guide/wdcguide_a.html.
2. Ibid.
3. Joseph M. Davila, Arthur I. Poland, and Richard Harrison, "International Heliophysical Year: A Program of Global Research Continuing the Tradition of Previous International Years," *Advances in Space Research* 34, no. 11 (2004):

2453–58, https://www.sciencedirect.com/science/article/abs/pii/S02731177 04006398.

4. "New Portrait of Our Planet: What IGY Taught Us," *Life*, November 7, 1960.

5. Interview of Nathaniel Gerson by Fae Korsmo, April 11, 2001, Niels Bohr Library & Archives, American Institute of Physics, College Park, MD, www.aip.org/history-programs/niels-bohr-library/oral-histories/24331.

6. Ibid.

7. Interview of Herbert Friedman by Martin Harwit, June 7, 1983, Niels Bohr Library & Archives, American Institute of Physics, College Park, MD, www.aip.org/history-programs/niels-bohr-library/oral-histories/28186.

8. Ibid.

9. Ibid.

10. Ibid.

11. Ibid.

12. Ronald Fraser, *Once Round the Sun* (New York: Macmillan, 1961), 159.

13. Walter Sullivan, *Assault on the Unknown* (New York: McGraw-Hill, 1961), 417.

14. Alexander Marshack, *The World in Space* (New York: Dell, 1958), 189.

15. Sydney Chapman, *IGY: Year of Discovery* (Ann Arbor: University of Michigan Press, 1959), 109.

16. J. Tuzo Wilson, *IGY, the Year of the New Moons* (New York: Alfred A. Knopf, 1961), 329.

17. Interview of Artjob Povzner and Lilia Morozovskia by Tanya Levin, January 19, 1999, Niels Bohr Library & Archives, American Institute of Physics, College Park, MD, www.aip.org/history-programs/niels-bohr-library/oral-histories/33699-1.

Index

active regions, 47, 58
Aerobee, 160
aging, Soviet Union and, 152
Ahlmann, Hans, 185
airglow, 7, 105
albedo, 186–89
all-sky camera, 105–6
altimetry, Antarctic research and, 82
American Astronomical Society, 188
American Institute of Physics, 23, 79, 129
American Meteorological Society, 178
American Society for the Advancement of Science, 67
Amundsen-Scott South Pole Station, 74
Antarctic research, 25, 71–88; and aurora, 95; Berkman on, 31, 33; and carbon dioxide levels, 82–83, 130, 131–32; and continental drift, 120; and elevation, 80–82; Huffman on, 89; personal experience of, 71–72; publicity and, 24; results in, 86–88; Van Allen and, 159
Antarctic Treaty, 34, 84–86, 89, 152
Antarctic Weather Central, 189
anthropogenic effects: and climate, 182, 184; and water cycle, 181–82
apex leadership, Nichols on, 211
Apollo 8, 165
Appalachian Mountains, 116, 124

Applied Physics Laboratory, Johns Hopkins, 158, 159–60
Arctic research, 24; and aurora, 95; Berkman on, 31; and global warming, 185; Van Allen and, 159
areas of inquiry, 7
Argentine Navy, 127
art: aurora and, 96; Goins on, 112
astrobiology, Soviet Union and, 151
astrolabe, 187
Atacama Large Millimeter Array, 153
Atlantic Ocean survey, 207–8
ATLANTIS–DISCOVERY I, 208
Atlas, 136
atmospheric pressure, physical effects of, 2
atmospheric research: IGY and, 5; interdisciplinary, 19; Second Polar Year and, 4
Atoms for Peace, 34
audacity, Nicolet on, 30
aurora(s), 7, 42, 93–107, 141; Antarctic research and, 74, 95; australis, 48; borealis, 2, 48, 95; cause of, 99–100; effects of, 40, 98; personal experience of, 93–94, 177; polaris, 96; results of research projects on, 105–7; study of, rationale for, 103
aviation industry: Antarctic and, 76; space weather and, 55

Baker-Nunn astronomical tracking cameras, 140, 144
Bentley, Charles, 32, 78–83
Berkman, Paul Arthur, 31–36
Berkner, Lloyd V., 5–7, 17–18, 20–21, 25, 74, 157, 158
Bifrost Bridge, 97
billionaires, and oceanography, 210
biology, 151, 206–7
Boulder Observatory, 46, 59
Browning, Elizabeth Barrett, 23
Bush, George H. W., 195
Bush, George W., 64
Byrd, Richard E., 17
Byrd Station, 74–75

carbon dioxide levels: Antarctic research and, 82–83, 130, 131–32; Keeling and, 129–30, 199–202; measurement of, 190, 193, 208
Carnegie Institute, 159
Carrington, Richard, 97
Carrington Event, 48, 97–99
Center for History of Physics, American Institute of Physics, 129
Center for Science Futures, 112
Central Radio Propagation Laboratory, 60–61
Chapman, Sydney: on aftermath, 220; and aurora, 97, 101–3; and China, 27, 28; and planning, 5–8, 17–18, 26, 157–58
Charbonneau, Rebecca, 149–55
Chile, 153
China, 26–29, 110
Christmas Island, 168
chromosphere, 42
Chu Chia-hua, 27–28
citizen diplomacy, 153
climate, versus weather, 182
climate change, 183–87; Berkman on, 32; history of, Antarctic research and, 76–78; Townley on, 13–14, 16. See also contemporary collaboration; global warming

climate cycle, Berkman on, 35
climate education, Niepold and, 63–69
climate literacy, term, 65–66
Climate of HOPE, 14, 89–90
CMEs. See coronal mass ejections
Cold War, 118, 185–86, 200; Charbonneau on, 151–54; Morozovskia on, 221; Paulavets on, 172
collaboration: in Antarctic research, 75–76; Antarctic Treaty and, 84; Berkman on, 33; Berkner and Chapman and, 17–18; Charbonneau on, 152; Friedman on, 218; Gerson on, 216–17; Goins on, 113; Nichols on, 209–10; Niepold on, 65–69; Paulavets on, 170; Townley on, 14; Weart on, 131; in weather science, ix; Wilson on, 9–10. See also contemporary collaboration
communications, 7; auroras and, 98–99; Townley on, 15–16
computers, and data processing: and aurora research, 102, 107; and meteorology, 181; and solar research, 61
conjugate point observations, of auroras, 104
contemporary collaboration on climate change, possibility of, 11; Huffman on, 89; Niepold on, 67–69; Paulavets on, 172–73; Shepherd on, 195–96; Townley on, 13
continental drift, 90, 115–28, 201
Cornell University, 60, 74, 107
corona, 43, 45; Mason on, 47–48
coronagraph, 57
coronal mass ejections (CMEs), 2, 41–42; French on, 50–55; indirect detection of, 59; line profiles of, 57–59; magnitude of, 51; patrol stations, 60; and satellites, 49
cosmic rays, 7; Antarctic research and, 74–75, 87; astronauts and, 165; satellites and, 145–46; Van Allen and, 159–65

courage, Nicolet on, 30
COVID-19, 196
Craig, Harmon, 200
crises, and data management, 110, 112
Cromwell current, 205
crustal structure, 125, 127–28
CSAGI, 7, 19–21, 61
cuped bay, 87
curriculum, Townley on, 12, 14
cybernetics, Soviet Union and, 152

Dalrymple, Brent, 117, 123–25
Danjon, André-Louis, 187
Danjon Prismatic Astrolabe, 187–88
data access, 5, 8, 14, 29; Antarctic Treaty
    and, 85; Niepold on, 68; and satellite
    programs, 137–39; Weart on, 131
data management/processing: aftermath
    and, 215–16; and aurora research, 101,
    107; and carbon dioxide measurement,
    199–200; Charbonneau on, 149; and
    continental drift, 121, 124; Goins on,
    109–14; and meteorology, 189–90;
    and oceanography, 197; planning
    for, 8, 26; and satellites, 140–41; and
    seismology, 126; and solar research,
    60–61
Davila, Joseph M., 216
decarbonization, Niepold on, 68
Dietz, Robert, 123
Digital Object Identifier, 112
diplomacy. See science diplomacy
diversity: and genetic research, 114;
    Huffman on, 91
Doel, Ronald E., 120, 122, 123, 185–86
DOLDRUM, 205
DOLPHIN, 204–5
DOWNWIND, 204
Drake, Charles, 125
Dryden, Hugh L., 19, 20

Earth: albedo of, 186–89; formation of,
    37; interdisciplinary research and, 19;
    magnetic field of, 55

earthquakes, 125
earth science, 115–28; Niepold on, 63,
    68–69
Easter Island Rise, 127, 128
education. See science education
Eisenhower, Dwight D., 33–34, 135,
    152, 217
Ellsworth Station, 75
El Niño, 178, 180
epochal observations, 9
Ericson, D. B., 207
European Southern Observatory, 153
European Space Agency, 218
exchange programs: Charbonneau on,
    153–54; Wilson on, 9–10
Explorer program, 146, 162, 166

Fanning Island, 168
Fifth International Polar Year, proposed,
    11, 113
Finck, W. Arnold, 198–99
First Polar Year, 3–4, 8; and auroras, 95,
    100; Berkman on, 35
Ford Foundation, 22
forecasted events, observations of, 25
fossil fuel industry, 89, 90, 201
Frankel, Henry, 118–19
Fraser, Ronald, 17–18, 184, 219
French, Ryan, 50–56, 98
Friedman, Herbert, 105, 142, 161,
    217–19
funding issues: Charbonneau on, 149–50;
    and data management, 112–13; Finck
    on, 198; Friedman on, 217; and
    instrument production, 46; Keeling
    and, 130, 199; Paulavets on, 170, 173;
    and planning, 19, 21; and satellites,
    136

Galilei, Galileo, 43, 95
Garbage Queen, 16
Gartlein, Carl, 102
geochemical measurements, 193
geographical areas, 9

geomagnetism, 2, 7, 20; Antarctic research and, 74–75, 87; results in, 166–68; solar radiation and, 49

geophysics, 206–7; Dryden on, 20; and ocean circulation, 204–5, 207–8

Geophysics Laboratory, University of Toronto, 121

gerontology, Soviet Union and, 152

Gerson, Nathaniel, 216–17

glaciology, 7; Antarctic research and, 80, 83, 87–88; Berkman on, 31–32

Global Change Act, 63

global warming, awareness of, 77, 82–83; *Life* and, 183–84, 185; Nichols and, 211; Revelle and, 200–202; Weart and, 129–32; Wilson and, 191. *See also* climate change

Goins, Meredith, 109–14

gravity, 7; Antarctic research and, 82, 88

greenhouse effect, 199

guano industry, 185

Gulf Stream, 180

Hagerty, James, 135

Hallett Station, 75

Harrison, Richard, 216

Hayes, E. Nelson, 140

helium, 38

Hess, Henry, 119

High Altitude Observatory, 38, 57–60

higher education, Townley on, 15

history of science: Charbonneau on, 149–55; Goins on, 111; Townley on, 14–16

hope: Berkman on, 35; Nichols on, 213; Niepold on, 68; Shepherd on, 196; Townley on, 13, 16

hotspots, 121

Huffman, Louise, 72, 77–78, 89–91

Hunkins, Kenneth, 23–25

hurricanes, 180, 181; forecasting storm surge, 206; Idalia, 1–2

hydrogen, 38

Hynek, J. Allen, 139–40, 141, 143, 187–88

ice research, 76–78, 87–88; and albedo, 186–87; and carbon dioxide levels, 131–32; Huffman on, 91; Hunkins and, 23–24; ice sheet thickness, 32, 184

IGY. *See* International Geophysical Year

IGY Bulletin, 22

indigenous knowledge, 12; Berkman on, 31; Huffman on, 91; Niepold on, 69

informed decision making, Berkman on, 32, 35

infrared rays, solar, 44

institutional memory, 150–51

instrument production, 17, 46; and auroras, 100–101, 104–7; Charbonneau and, 149, 154; Danjon and, 187–88; Keeling and, 199; and satellites, 140; and solar research, 39, 57–59; Van Allen and, 160

intercontinental ballistic missile (ICBM), 136

interdisciplinary research, 18–19; Huffman on, 90–91; Paulavets on, 171–72; Shepherd on, 195

Intergovernmental Panel on Climate Change, 195

international, term, viii

International Arctic Science Committee, 11

International Council of Scientific Unions, 18, 109

International Geophysical Year (IGY), vii–ix; aftermath of, 215–21; benefits of, 19, 53, 129–30, 216–18; context of, 15–16, 32; enthusiasm for, 9–10, 20–22, 30, 221; importance of, 31, 150; memories of, vii–viii, 47–48; mission of, 6–7; planning of, 17–30; predecessors to, ix–x; program criteria for, 8–9; proposals for, 5–10; timing of, 6–7; Van Allen and, 160

International Heliophysical Year, proposed, 216

International Science Council (ISC), 109, 111, 112; Paulavets and, 169–73

ionization, 45, 49–50, 59
ionosphere, 5, 7; Antarctic research and, 74
isolation/siloing of researchers: Paulavets
    on, 171; Shepherd on, 196. *See
    also* collaboration; contemporary
    collaboration

Japan, 153, 218
Jarvis Island, 168
jet stream, 206
Johns Hopkins Applied Physics
    Laboratory, 158, 159–60
Joyce, Wallace, 157, 158

K-coronameter, 57–58
Keeling, Charles David, 129–30, 199–200
Keeling Curve, 129
King, Joseph E., 204–5
Knauss, John A., 204–5

Lamont-Doherty Earth Observatory, 119,
    126, 198, 205–6
Landsat mission, 63, 68
La Niña, 180
leadership: Nichols on, 211; Odishaw
    and, 217
Leonard, Arthur S., 145
Leopold, Frank, 89–90
Life magazine, 216; on Antarctic, 74,
    76; on continental drift, 120–21; on
    global warming, 183–84, 185; on
    meteorology, 180; "New Portrait of
    Our Planet," 44, 183–84; on solar
    research, 44; on sunspots, 42–43; on
    Van Allen radiation belts, 164
Lightfoot, Robert, 150
Little America Station, 74
Livy, 96
longitudes/latitudes, 7
Long Playing Rocket, 134, 161
Ludwig, George H., 166
Lyot telescope, 39

magnetosphere research, 166–68, 218
mantle, 117, 119–20, 121

marine biology, 206–7, 209–10
marine geology, 119
Mariner II, 54
Mars, 54–55
Marshack, Alexander, 26; on aftermath,
    220; on auroras, 42, 95–96, 99, 101–
    2; on global warming, 185; on solar
    research, 43–44, 46
Mason, Emily, 47–50
Mauna Loa Observatory, 199
Maury, Matthew, ix–x
McMath-Hulbert Observatory, 46, 59–60
McMurdo Naval Air Facility, 75
medicine, Soviet Union and, 151
Memel patrol spectrograph, 104, 106–7
Mercury, albedo of, 187
meteorology, 1–10, 182; Antarctic
    research and, 74; and collaboration,
    ix; early projects in, ix–x; forecasting
    storm surge, 206; NOAA and,
    178–80; oceanography and, 197–98;
    planned synoptic observations of, 26;
    results in, 192–93; satellites and, 146–
    47; Shepherd on, 177–78, 180–81;
    Van Allen and, 159
military issues: Antarctic Treaty and, 84–85;
    auroras and, 99; First Polar Year and, 3;
    ice cap thickness and, 185–86; Paulavets
    on, 170; rockets/satellites and, 133–47;
    seafloor mapping and, 210
misinformation industry: Huffman on,
    89–90; Townley on, 14–15
mission science: Niepold on, 63–65;
    Paulavets on, 169–73
Moonwatchers, 139–40, 144–45
Morley, Larry, 122–23
Morozovskia, Lilia, 221
Mount Wilson Observatory, 46
mythology, and auroras, 96–97

National Academy of Sciences, 6, 18
National Aeronautics and Space
    Administration (NASA): on albedo,
    187; Friedman and, 217–18; Niepold
    and, 63–65; Shepherd and, 176–78

National Astronomical Observatory of
  Japan, 153
National Bureau of Standards, 38–39,
  60–61
National Center for Science Education,
  11–16, 90
National Climate Assessment, 195
National Institutes of Health, 114
National Institution of Oceanography
  (England), 126
National Oceanic and Atmospheric
  Administration (NOAA), 51, 63–69,
  110, 178–79
National Radio Astronomy Observatory,
  149–55
National Research Council, 18
National Science Foundation, 18, 23,
  150
native form, 197
neuropsychology, Nichols on, 211–12
Nichols, Wallace J., 179, 209–13
Nicolet, Marcel, 20, 24, 27–28, 30
Niepold, Frank, 63–69
NOAA. *See* National Oceanic and
  Atmospheric Administration
northern lights. *See* aurora(s)
nuclear power, Antarctic Treaty, 86
nuclear radiation, 7; program planning
  for, 20
nuclear weapons: Berkman on, 33–34;
  satellites and, 142–43; tests, and ocean
  waves, 203, 206

Oak Ridge Innovation Institute, 109
observation, 17; Antarctic Treaty and,
  85; and auroras, 95, 100, 103–6;
  Berkman on, 35; Dryden on, 20; and
  earth albedo, 187; magnetic, 166–68;
  and meteorology, 189; of sea levels,
  197; upper air, 192–93; Van Allen
  and, 162–65; volunteers and, 64, 101,
  139–41, 144–45
ocean(s): age of, 208; carbon dioxide
  capacity of, 129–30, 200–201;
  circulation of, 204–5, 207–8;

equatorial countercurrent, 205; floor
  of, heatflow, 204
Ocean Literacy Guide, 65
oceanography, 7, 197–208; Antarctic
  research and, 75; interdisciplinary
  research and, 19; Nichols on, 209–13;
  NOAA and, 178–80; ports permission
  and, 130–31; program planning for,
  19–20; results in, 202–8
O'Connell, John "Gene," 71
Odishaw, Hugh, ix, 217
Office of Naval Research, 60, 135
O'Keefe, John A., 120
open science, Goins on, 113–14
Open Skies, 33–34; Charbonneau on,
  149–50
optical telescope, 39
Ordway, Frederick, 33
The Outer Limits, 97
outer space exploration, Huffman on, 90
Outer Space Treaty: Berkman on, 34;
  Charbonneau on, 152
oxygen, 38
ozone layer, 45; measurement of, 190, 193

Palmyra Island, 168
Pangaea, 115–16, 120
paradigm shift, continental drift and,
  115–28
Paris Agreement, 68
Parker, Eugene, 53–54
Parker Solar Probe, 54
partnerships: Niepold on, 65–69. *See also*
  collaboration
Paulavets, Katsia, 169–73
peer review, 110
Petruccelli, Antonio, 183
photosphere, 42, 53
Planet Earth films, 22
plankton, 185
plate tectonics. *See* continental drift
Poland, Arthur I., 216
political issues: Antarctic Treaty and, 85;
  Berkman on, 35; China/Taiwan and,
  26–29; planning and, 21

Povzner, Artjob, 197–98
power grids: auroras and, 98–99; solar radiation and, 52
promotional efforts, 22–24; and Antarctic, 83; Charbonneau on, 153; solar research and, 42; space and, 133

radar, and aurora research, 103
radiation belts. *See* Van Allen radiation belts
radio: atmosphere and, 4, 45; aurora and, 104; and satellite tracking, 138; and Second Polar Year, 4; solar radiation and, 42, 44, 49, 51, 60
Radio Astronomy Station, Fort Davis, Texas, 46
radio proximity fuze, 159
random events, observations of, 25
Rawinsonde observations, 192–93
Ray, Ernie, 166
Regular World Days (RWDs), 25–26
relationships, Charbonneau on, 155
Reporters, 7
research directions: polar, 74–75; solar, 41, 45–46; space, 134, 140–41
Revelle, Roger, 129, 200–202
rockets, 7, 33–34, 133–47, 160; and aurora research, 101–2; Friedman on, 218; Hunkins on, 24; program planning for, 20; and solar research, 42
rockoons, 40
rugged vacuum tube, 159
Russia: Berkman on, 35. *See also* Soviet Union

Sacramento Peak Observatory, 42, 46, 59–60
salmon industry, 185
satellite(s), 7, 40, 133–47; and albedo research, 188; Friedman on, 217–18; solar radiation and, 49, 51–52; space race and, 29; and space weather, 176; tracking, 138–40, 143–44; Weart on, 130

Satellite Panel, 161
science diplomacy, 12; Berkman on, 31–36; Charbonneau on, 149–50, 153–55
science education: goals of, 12; Huffman on, 89–90; IGY promotion and, 22–23; Niepold and, 63–69; Shepherd on, 196; Townley on, 11–16
Scientific Committee on Antarctic Research, 11
Scott Expedition, 75
Scripps Institute of Oceanography, 74, 119, 127, 168, 200–201, 204
seafloor: age of, 201; mapping, 210; spreading of, 119–20, 122, 124–25
sea levels, 202–4, 205–6; Antarctic and, 80–83, 88; climate change and, 184; observation of, 197
sea turtles, 209
sea water, radiochemical analysis of, 208
Second Polar Year, 4, 7; scope of, 7–8
seismology, 7, 115–28; Antarctic research and, 75, 80, 82; results in, 125–28
Sfraga, Mike, 23
Shapley, A. H., 61
Shepherd, James Marshall, 176–78, 179–81, 195–96
Shoemaker, Brian, 157–58
Singer, Fred, 157
Smithsonian Astrophysical Observatory, 144–45
snow cat parties, 25, 78–79, 82; term, 23
social sciences, 12; Berkman on, 31; Paulavets on, 171–72
Sokolsky, Steve, vii–viii
solar flares. *See* coronal mass ejections
solar maximum, 51
solar prominences, 41
solar research, 7, 37–61; interdisciplinary, 19; results in, 57–61
solar wind, 40, 53–54; Mason on, 48–49
South Africa, 98, 190
South American Meteorological Services, 192
Soviet Union: Berkman on, 33; Charbonneau on, 151–53; and

data access, 151; and data centers, 8; and ice cap thickness, 185–86; and magnetosphere, 218; and oceanography, 197–98; and satellites, 29, 137–39, 141–43

space medicine, Soviet Union and, 151

space race, 139, 141–43; Charbonneau on, 152–53

space research, 133–47, 157–68; results in, 143–47, 166–68

space weather, 175–93; app, 50; French on, 50–55

Sputnik, 33–34, 139, 141–43, 152

standardization: Antarctic research and, 88; as goal, 17; Goins on, 111; National Bureau of Standards and, 39; Niepold on, 64, 68–69

Starlink, 49

Sternberg Astronomical Institute, 151

St. Onge, Tim, x

Stratospheric Analysis Project, 193

sublimity, Nicolet on, 30

Suess, Hans, 201

Sullivan, Walter: on aftermath, 219–20; on Antarctic Treaty, 85–86; on Arctic, 24; on auroras, 101, 102; on climate change, 184–85; on cosmic rays, 163; on criteria, 8; on political issues, 27, 29; on rockets, 160; on satellites, 138; on timespan, 7; on weather, ix, 4

sun, 37–61; atmosphere of, 42, 44–45, 48; characteristics of, 37–38; French on, 50–55; layers of, 42–43; Mason on, 47–50; rotation of, 43. See also under solar

sunspots, 4, 41; Carrington and, 97; French on, 53; Galileo and, 43; IGY timing and, 7, 42

sustainability: and data management, 111–12; Paulavets on, 169–73

sustainable development goals (SDGs), 170–71

Sykes, Lynn, 122, 123

synoptic observations, 9; Friedman on, 218; and oceanography, 198;

planning, 25–26, 61; and solar research, 43–44, 59–60

Taiwan, 26–29

teachers: IGY educational materials and, 22–23; Townley on, 12, 15

telegraph wires, auroras and, 98

telescopes: Charbonneau on, 149–50, 153, 154–55; Galileo and, 43; optical, 39

Townley, Amanda, 11–16, 90, 176

transform faults, 121–22

traverses, 78–82, 126

tsunami, record of, 203

ultraviolet waves, solar, 44–45

United Nations Educational, Scientific and Cultural Organization, 21

United Nations Institute for Training and Research, 31

University Corporation for Atmospheric Research, 186

University of Alaska Geophysical Institute, 24, 100

University of Colorado High Altitude Observatory, 38

University of Georgia, 178, 195

University of Hawaii, 46, 60

University of Michigan at Ann Arbor, 46

University of Tennessee, 109

University of Toronto Geophysics Laboratory, 121

University of Victoria, 109

University of Washington, 208

University of Wisconsin–Madison, 80, 82

upper atmosphere: Antarctic research and, 87; interdisciplinary research and, 19; physics of, 141

U.S. Army Signal Research and Development Laboratory, 146

U.S. Coast and Geodetic Survey, 123, 167

U.S. Geological Survey, 74, 116

U.S. Ice Drilling Program, 72

U.S. Navy, 135–36

V-2 rockets, 135, 160
Van Allen, Abigail, 157, 158
Van Allen, James, 157–66
Van Allen radiation belts, 157, 161–66;
    Antarctic research and, 75; Friedman
    on, 218; Huffman on, 90
Van Biesbroeck, G. A., 187
Vanguard, 136–37, 140–42, 145, 146
Vema, 198–99, 206
Venkatesan, Doraswamy, 162
Venus, albedo of, 187
very long baseline interferometry
    technique, 154
Vestive, E. H., 157
Viking, 135–36
volunteers: Moonwatchers, 139–40, 144–
    45; and observation, 64, 101, 141
von Braun, Werner, 33–34, 135
Vostok, 131
Voyage to the Bottom of the Sea, 162

water, Nichols on, 212
water cycle, 181–82, 191
Weart, Spencer, 129–32, 190, 199
weather. *See* meteorology; space weather
weather balloons, 193
Weather Bureau, 190, 192, 193
Weekly Reader, viii, 90
Wegener, Alfred, 116, 119, 120, 123, 124
West, Gordon F., 119, 122–23
Wexler, Harry, 129–30
Weyprecht, Karl, 3
Whipple, Fred, 134, 139–40, 141

Whistler waves, 75
white-light corona photometer, 57–58
Wilkes Station, 75
Willard, Ted, 67
Wilson, J. Tuzo "Jock": and aftermath,
    220–21; and Antarctic, 73; and
    aurora, 96, 101; and continental drift,
    118–23; and exchange, 9–10; and
    planning, 6; and solar wind, 40; and
    sun, 38, 39; and water cycle, 191
Windridge, Melanie, 96–97, 99
Woods Hole Oceanographic Installation,
    126, 208
working groups: and Antarctic research,
    74; organization of, 19–20
World Data Centers (WDCs), 8, 109,
    190, 204; aftermath and, 215–16
World Data System (WDS), 109–14
World Days, 7, 25–26, 103
World Meteorological Intervals (WMIs),
    25–26
World War II, 118, 133, 185
World Warning Agency, 60
World Warning Center, 42

X-rays, solar, 45

year, term, viii–ix, 7
Yerkes Observatory, 106
youth: Niepold on, 68; Shepherd on,
    196; Townley on, 13, 16

zoology, Antarctic research and, 75